SOTHIS PRESS

Sothis Press aims to completely reimagine the public domain book with an emphasis on universal access to captivatingly designed print editions. We believe that technology will enable every single book ever published to be perpetually accessible in beautiful print form. Sothis is beauty in ubiquity.

www.sothis.press
1 Waterhouse Square, London, ECIN 2ST

Boiler Feed Water

A Concise Handbook of Water for Boiler Feeding Purposes (Its Effects, Treatment, and Analysis)

Percy George Jackson

SOTHIS PRESS

ISBN 978-1-0368-0123-6

This edition is published by
Sothis Press © 2023. The company
is registered in England and Wales.
The Company Number is 14900389.

BOILER FEED WATER

A CONCISE HANDBOOK OF WATER FOR BOILER FEEDING PURPOSES

(ITS EFFECTS, TREATMENT, AND ANALYSIS)

BY

PERCY G. JACKSON, F.I.C.

ANALYTICAL AND CONSULTING CHEMIST; FORMERLY CHEMIST TO THE NATIONAL BOILER AND GENERAL INSURANCE COMPANY, LIMITED

ILLUSTRATED

SECOND EDITION, REVISED AND ENLARGED

LONDON

CHARLES GRIFFIN AND CO. LIMITED

EXETER STREET, STRAND, W.C. 2

1922

PREFACE TO SECOND EDITION

THE reception given to the first edition has been most gratifying, and confirms my idea that there was an opening for such a work, particularly among those, both chemists and engineers, who have to assist in the control of boiler plant, but who have not the time, or the chemical knowledge necessary, to make a special study of the subject.

Suggestions have been made that the book should be considerably enlarged to make it a complete and up-to-date work of reference on the subject, going into a full discussion of the subject of corrosion, and giving detailed descriptions of all the principal makes of softening plant. I have decided not to adopt the suggestion, as the question of corrosion has been so ably dealt with by other writers, and detailed descriptions of softening plants are furnished by the makers.

At the same time I recognise that the chapter on softening plants could very well be made more useful by outlining the special features of the plants of the best known makers, and I have, therefore, entirely revised the chapter, and trust it will now give sufficient information to enable anyone to make a preliminary selection when requiring to install such a plant.

The advantages of softening water for use in laundries is, I think, outside the scope of the book. Perhaps it may suffice to point out here that one thousand gallons of water with ten degrees of hardness will destroy about fourteen pounds of genuine soap. As the cost of lime-soda réagent for removing ten degrees of hardness would be less than one penny per thousand gallons of water, the conclusion is obvious.

The character of the book has not been materially altered, but a chapter has been introduced dealing with boiler control, and a few additions have been made to the chapter dealing with corrosion. Methods for determining oil in water and in scale, formulæ for the calculation of the number of days a boiler may be worked before reaching a desired density, and a table giving the capacities of horizontal cylinders filled to different depths have been added. The question of the continuous blowing down of boilers has been referred to, and also the removal of oxygen from feed water.

I wish to thank all those who have offered any criticism or made suggestions for the improvement of the book, as well as the various makers of softening plants for permission to use their illustrations and for the loan of blocks.

I trust that the new edition will be as well received as the first, and I shall always be pleased to hear from readers who have suggestions likely to be useful in any future editions.

PERCY G. JACKSON.

13 St. Clement's Road,
Chorlton cum Hardy,
MANCHESTER, *May*, 1922.

PREFACE TO FIRST EDITION

So far as I am aware no book on the subject of Boiler Feed Water has hitherto been written by any one having extensive experience as chemist to a boiler insurance company, and I feel that an intimate practical knowledge of the subject cannot otherwise be so well obtained.

The investigation of occasional isolated cases of feed water trouble is undoubtedly very useful. and much helpful information is thus obtained, though it can hardly be expected to yield the same general knowledge of the subject as that resulting from the consideration of analyses in conjunction with the history of the boilers, or the batteries of boilers concerned, though I am well aware that even then there are many possible factors which are not thus disclosed. Slight variations in the working conditions, which may not even be brought to notice by a special investigation of a particular case, can yet have an important bearing on the subject, especially in those instances where one or more boilers in a battery are affected, while the others are in a satisfactory condition.

Makers of water softening plants have not usually the expert chemical staff necessary thoroughly to investigate troubles which continue even after their plants have been put to work, nor should they be (although they usually are) expected to render continued and expensive services gratis, after having put down a plant which is capable of doing its work efficiently. Some of them do keep in touch with their plants for a long time, and offer regular advice as to necessary changes in the treatment as judged from log sheets of daily tests made by the users, but beyond this they cannot reasonably be expected to go. If they erect a plant capable of reducing the hardness-forming salts to a minimum, and eliminating oil; and if the plant is so designed and constructed as to continue to work satisfactorily with reasonable attention for a number of years, they may be considered to have fulfilled their obligations, unless, of course, they have given special guarantees to effect specified results.

It is not intended that this book shall explain theoretically why various constituents of feed water cause certain undesirable results, but only how experience has shown that these undesirable results may be eliminated, or at least mitigated.

There is still much to be learnt on the subject, and probably will be until some one comes forward with unlimited time to devote to such matters, and who is unhindered in the thorough prosecution of his researches by financial considerations. He must not expect to carry through the work in a laboratory, but must spend much of his time in the stoke-hole, always observant of details, able to advise slight or important changes in the working conditions, and note the results obtained thereby.

All efforts to fathom boiler feed water troubles are

handicapped by the fact that it is impossible to see what is actually happening inside a boiler while it is at work, and by the impossibility of obtaining exact working conditions in laboratory experiments.

This work is the result of the experience largely gained by me while working as chemist to the National Boiler and General Insurance Company, Limited, and I wish to express my indebtedness, for allowing me to publish this book, to the Company, and to Mr. Edward G. Hiller, the chief engineer, and other engineers of his staff, who have during the years of my service given me the benefit of their experience.

I trust the information in this book will be found to be reliable, concisely stated, and of practical use.

PERCY G. JACKSON.

January, 1919.

CONTENTS

CHAPTER I

INTRODUCTORY

PERFECT boiler feed water can never be obtained from any natural source, and though a small proportion of the available supplies may be considered suitable for use without treatment, most of them require treatment in some way or other if the boilers using them are to be kept as clean and free from corrosion as possible. A few may be non-corrosive, but practically none can be found which do not form more or less deposit in boilers, and as a rule some of the deposit remains on the plates or tubes in the form of scale.

Unsuitable feed water in boilers may cause damage in a number of ways such as :—

(*a*) Corrosion of the plates.

(*b*) Corrosion of the fittings.

(*c*) Grooving of the angles and seams.

(*d*) Leakages, and sometimes shearing off of rivets owing to straining of the seams through excessive expansion and contraction of the plates.

(*e*) Bulging, hogging, or ripping of the tubes. Even the shell plates may be bulged.

(*f*) Engine troubles due to priming of the water in the boilers.

Corrosion of the plates of a boiler may occur in several forms, of which the principal are :—

(*a*) *Isolated pittings*, which may sometimes perforate a plate without materially reducing its strength.

(*b*) *Honeycomb pitting*, consisting of areas of close isolated pittings. This is more serious than the above, its importance depending on the size and depth of the pits and the intervals between them.

(*c*) *Confluent pitting*, which occurs in areas of varying size and may cause serious weakening of the plates.

(*d*) *Smooth wasting*, by which the whole of the original surface of the metal in a particular area is removed, leaving a more or less uniform surface, which may not be readily noticeable. The depth and extent of the wasting thus occurring is often difficult to determine, and serious weakening of the plates may have taken place before it is detected. When suspected it is usually necessary to drill the plate and gauge the remaining thickness.

(*e*) *Grooving* in angles and at seams, or at any junction of two plates. This may lead to the formation of cracks, or even to rupture.

(*f*) *Graphitic wasting* of cast iron. In this form a portion of the iron is apparently dissolved away, while the remainder is converted into black oxide, which, together with graphite and silica, remain behind in the form of, and similar in appearance to, the original metal, for which it can readily be mistaken. This form usually occurs in economiser pipes.

The causes of corrosion will be dealt with in another chapter.

Leakages, Bulging, etc., are usually caused by overheating of the tubes and plates of boilers mainly owing to scale and deposit, especially when grease is present. Scale in boilers not only retards the passage of heat through the plates or tubes, but also, by keeping the water from direct contact with the metal, may cause it to become more or less overheated, and result in undue strains being set up. It frequently happens, usually when grease is present in the scale, that the overheating is so severe as to lead to softening of the metal and consequent bulging or hogging of the plates or tubes, shearing of rivets and opening of seams, and possible collapse or rupture. Sometimes portions of tubes may be seriously burnt, presumably by becoming partially emptied of water, the metal then becoming red hot and combining with the oxygen in the water vapour. Such oxidation or burning may be of considerable depth. Even soft deposit, especially if of a light floury nature, or containing grease, may lead to serious overheating of the metal on which it is deposited, especially if it collects on those parts exposed to the greatest heat of the furnaces.

The overheating of plates caused through scale or deposit sometimes leads to corrosion by bringing about chemical changes in the constituents of the water, which might not have occurred at the lower normal temperature of the plates. Such decomposition will take place in actual contact with the plates, and the corrosive compounds formed, being shielded from the water by the scale, cannot readily be washed away, and may consequently continue their action for some time.

Grease is very objectionable in boilers, not only on account of the overheating troubles it may lead to, but also owing to the possible corrosion it may cause if it contains either animal or vegetable oil. Animal and vegetable oils contain fatty acids in combination with glycerine, and under the conditions which exist in steam boilers the fatty acids may become liberated and attack the plates.

Priming or Foaming is caused by certain constituents in the water in the boiler, especially where the conditions of working lead to sudden heavy demands on the steam. This trouble will be dealt with in a later chapter.

As previously stated, perfect feed water cannot be obtained from any natural source, and it may be fairly said never to be found at all, unless specially distilled for the purpose in evaporating plants such as are used in marine practice. The distilled water obtained as condensate from surface or other similar condensers almost always contains grease, and frequently mineral constituents carried over from the boilers in the steam. It is commonly considered that the condensate from steam turbines is free from grease, but in practice it is seldom found to be entirely so. Small quantities of grease, too small to affect the appearance of the water, and which may escape detection by simple tests, can yet have a very decided effect in causing overheating, especially in heavily fired boilers.

Passing greasy steam through oil separators will result in the removal of a large proportion of the grease, but there will always remain traces which cannot be removed by mechanical means.

If surface condenser water only could be used, little harm might result, but it is almost invariably necessary to introduce a certain proportion of make-up water, and this will in all probability contain some scale-forming constituents, which, when precipitated in the boilers, will entangle a portion of the oil, and result in a greasy scale or deposit. It may be thought that if only a small proportion of make-up is needed, or if it contains little scale-forming matter, it will be of no consequence. This is a mistaken idea, as with a given proportion of oil, the less the amount of deposit formed, the larger will be the percentage of oil in it, and the power of preventing the water coming into contact with the plates thereby increased.

It is surprising how little grease may have serious consequences. One part of grease in a thousand parts of scale or deposit may lead to bulging or hogging in heavily fired boilers.

Rain water is often responsible for corrosion in boilers owing to traces of acid gases dissolved from the atmosphere, especially when collected in large towns or manufacturing districts, and it usually contains a certain amount of suspended matter composed of soot and dust. At times rain water contains appreciable quantities of mineral matter in solution even if collected direct into a clean vessel. As much as a hundred degrees of hardness has been found in samples thus carefully collected.

Speaking in general, well waters are usually the most sedimentary, and also more likely to contain corrosive mineral constituents than water obtained from other sources. River waters, as a rule, have less scale-forming salts, and unless polluted are not so likely to be as corrosive as well supplies. Surface waters collected in moorland districts are mostly of a soft nature, but owing to peaty matter which they usually contain, and their comparative freedom from calcium carbonate, may have decided corrosive properties.

The *Hardness* of water means its power of destroying soap by the displacement of the sodium or potassium of the soap by calcium, magnesium, or other metals, which form insoluble salts with fatty acids, and are, therefore, precipitated.

Throughout this book *Degrees of Hardness* means English degrees—that is, the equivalent in grains per gallon of calcium carbonate, of the total calcium and magnesium salts present. Where iron, aluminium, or manganese salts are present in a water, their equivalent in calcium carbonate should also be included in the *Hardness*, as they, too, have soap-destroying properties.

The term " *Temporary Hardness* " was originally applied to that portion of the hardness which was precipitated by boiling. This is principally the carbonates, but as some of the carbonates will still remain in solution after even prolonged boiling, while some of the other salts may be precipitated in addition to carbonates, the hardness thus represented is somewhat indefinite and variable, depending to a certain extent on the exact method of determination, as well as on the other salts present in the water. It is preferable therefore to consider " *Temporary Hardness* " to be all the hardness present in the form of carbonates, and the term is thus used in this book.

" *Permanent Hardness* " is used to indicate all hardness present other than as carbonates.

" *Alkalinity* " is the equivalent, in terms of grains per gallon of calcium carbonate, of those salts which have the power of neutralising acids. This is usually the carbonates, though in waters which have been chemically treated it will include caustic lime and caustic soda if present. It is sometimes referred to as " *Degrees of Alkalinity* " for convenience, as it is comparable with " *Degrees of Hardness.*"

CHAPTER II

MINERAL CONSTITUENTS

THOUGH waters differ widely in constitution, the mineral salts usually present are comparatively few, comprising the carbonates, sulphates, chlorides, and nitrates of calcium magnesium, and sodium, together with a small proportion of silica. Potassium is probably present as a rule, but it is seldom considered necessary to test for or estimate it, and it is therefore usually included with the sodium salts when reporting results. Some waters also contain appreciable, and occasionally large, quantities of the salts of iron, aluminium, and manganese.

It is usual, when discussing the composition of water, to speak of *carbonates*, but strictly speaking this is incorrect in most cases. In natural waters *carbonates* are always present in the form of *bicarbonates*, which compounds contain twice as much carbonic acid (carbon dioxide) as the carbonates. The additional carbonic acid required to convert carbonates into bicarbonates is only loosely held, as mentioned later, and may be conveniently referred to as "*Half-combined Carbonic Acid.*"

In the following discussion of the chief characteristics of the principal salts, the word "scale" may sometimes be intended to mean any deposit formed in the boilers, whether as actual scale, loose deposit, or mud.

Calcium Carbonate is derived from chalk or limestone and is the most common of all the scale-forming constituents found in waters. It is only sparingly soluble in water except in the form of bicarbonate, in which form, as previously mentioned, it is always present when found in natural waters. The solubility of the normal carbonate varies according to the nature and amount of other salts present, but it is usually considered to be soluble to the extent of about three grains per gallon in pure cold water. It is difficult to say what is the solubility of the bicarbonate, but twenty to thirty grains per gallon are found fairly frequently in very hard waters. Calcium carbonate is not only non-corrosive in boilers, but it has a tendency to neutralise the corrosive properties of other constituents, owing to its power of neutralising the acid is produced by their decomposition.

Calcium *bi*carbonate is readily decomposed by heat with the liberation of the half-combined carbonic acid, leaving the less soluble carbonate, and this results in its precipitation in boilers. It is similarly precipitated from solution by the addition of lime, which has a great affinity for, and therefore removes, the half-combined carbonic acid. This fact has led to the use of lime for the precipitation of calcium carbonate in softening plants.

Calcium Sulphate, usually derived from gypsum, is frequently present in water. It is fairly freely soluble in cold water, and its solubility increases slightly up to about 100° Fahr., at which temperature it is soluble to the extent of about one hundred and fifty grains per gallon. Above 100° Fahr. the solubility slowly decreases, while above 212° Fahr. it falls off rapidly. The solubility at the working temperature of most boilers is small, and as a result calcium sulphate is thrown out of solution in a crystalline form, and by itself will usually form a hard scale, somewhat resembling porcelain. On this account, calcium sulphate is a decidedly objectionable constituent in feed waters owing to the difficulty of removing the scale when cleaning the boilers. Calcium sulphate is considered to be an inert salt, and not liable, therefore, to cause corrosion in boilers.

On the addition of sodium carbonate (soda ash), calcium sulphate is precipitated as carbonate, and by this means it is usually removed in softening plants, sodium sulphate remaining in its place. Barium carbonate is occasionally used instead of soda ash for the precipitation of calcium sulphate owing to the fact that the salts resulting from the reaction (calcium carbonate and barium sulphate) are both precipitated, and the calcium sulphate is thus removed without leaving any other salt in its place. Unfortunately barium carbonate is so much more expensive than soda ash that the cost of treatment is usually considered prohibitive, otherwise it might be a most useful reagent in softening plants.

Calcium Chloride is only occasionally found in waters. Being extremely soluble in water at all temperatures, it is never likely to be deposited in boilers. It is a stable compound and not liable, therefore, to cause corrosion directly, but possibly it may interact with magnesium sulphate under suitable conditions, forming magnesium chloride, and thus leading to corrosion.

Calcium chloride is precipitated as carbonate by soda ash, leaving sodium chloride in solution.

Calcium Nitrate is also an extremely soluble salt occasionally present in water and never likely to be deposited in boilers. Like all nitrates it is probably liable to decomposition at high temperatures with the liberation of acid which will readily attack the boiler plates.

Calcium nitrate is precipitated as carbonate by soda ash, leaving sodium nitrate in its place.

Magnesium Carbonate, usually derived from magnesian limestones, is a fairly common constituent in water, and, like the corresponding calcium salt, is always present as bicarbonate. The bicarbonate is much more soluble in water than the carbonate, but even the normal carbonate is fairly soluble, decidedly more so than calcium carbonate. Magnesium carbonate is seldom found in the scale formed in boilers, owing to the fact that under boiler conditions it is almost invariably converted into hydrate. Though probably not directly corrosive, experience appears to show that, similarly to the other magnesium salts which occur in water, magnesium carbonate may interact with chlorides to form corrosive salts.

Like the corresponding calcium salt, magnesium *bi*-carbonate is converted into carbonate by heat, or by treatment with lime. The carbonate is, however, somewhat too soluble in water to be precipitated to any great extent, but if twice as much lime be used as is necessary to remove the half-combined carbonic acid, the additional lime further reacts with the carbonate to form the almost insoluble magnesium hydrate together with calcium carbonate, both of which will be precipitated. Caustic soda will similarly precipitate magnesium hydrate, but in this case sodium carbonate will remain in solution.

Magnesium Sulphate, commonly known as Epsom-salts, is found to some extent in most waters. It is a very soluble salt, but not nearly so soluble as the chlorides and nitrates of calcium and magnesium. Owing to its solubility, magnesium sulphate is not precipitated as such in boilers, and is therefore frequently classed as a non-scale-forming constituent. This is usually incorrect, as in practice it is seldom found to accumulate in boilers to any great extent. This is probably due to the fact that it may react, under boiler conditions, with chlorides and nitrates, the resulting salt being afterwards converted into the insoluble hydrate, and thus precipitated. Though experiments appear to show that magnesium sulphate is not materially corrosive, there is little doubt that in practice it usually reacts as above mentioned and forms corrosive salts, and it may safely be asserted that when a fair quantity of magnesium sulphate is present in conjunction with chlorides and nitrates of other metals, corrosion is likely to result.

Magnesium sulphate is usually removed in softening treatment by means of caustic soda, which precipitates magnesium hydrate, while sodium sulphate remains in solution. As a mixture of soda ash and lime will interact and form caustic soda, this mixture, being quite as effective and decidedly cheaper, is usually used in preference to caustic soda itself.

Magnesium Chloride, Magnesium Nitrate.—These are well known as very objectionable constituents in feed water. They do not usually occur in large proportion, but are frequently found in small quantities. Both are extremely soluble in water and are not, therefore, deposited in their original form, but by decomposition at boiler temperatures they are nearly always converted into the insoluble hydrate which is precipitated, while acids are liberated which readily attack boiler plates.

These constituents are removed in softening plants by caustic soda, or a mixture of lime and soda ash, in the same way as the sulphate.

Ferrous Sulphate (*Sulphate of Iron*) is the form in which iron usually occurs when present in quantity in water, and though in smaller quantities iron is frequently shown in analyses as carbonate, it is at least doubtful whether the carbonate is really present in solution as such.

Though sulphate of iron is fairly soluble in water, it usually commences to precipitate as basic carbonate or basic sulphate on exposure to air, especially if carbonates of other metals are present, and this precipitation generally continues until the water has become distinctly acid in reaction, provided sufficient sulphate of iron be present. In boilers, precipitation will take place more quickly, and a scale will be formed consisting of the basic iron salts.

When iron is deposited from water, either hot or cold, acid is liberated, and on this account waters containing it are extremely corrosive when used in boilers, and such waters should be avoided if possible for feeding purposes, and in any case should not be used unless efficiently purified by softening treatment.

Ferrous sulphate is removed in softening plants by caustic soda, or lime and soda ash, in the same way as magnesium salts.

Aluminium Sulphate is not so frequently found in quantity in waters as sulphate of iron, but it is similar to it in most respects, and no less objectionable, having extremely corrosive properties.

Manganese Sulphate is also occasionally present in fair proportion in water. Its effect is difficult to ascertain, as it is seldom, if ever, present except in conjunction with iron or aluminium. Probably it does little harm except to assist in the formation of scale.

Manganese sulphate will be removed in softening plants by the caustic soda, or lime and soda ash treatment.

Silica is usually found only in very small proportion, and is not of much importance, except that it slightly increases the quantity of scale. It may, however, when most of the other scale-forming salts have been removed by softening, produce a small quantity of very hard scale.

The above list comprises all the important scale-forming mineral salts commonly found in water, but the alkali metals, sodium and potassium, are almost invariably present in some form or other, such as bicarbonates, sulphates, chlorides, and nitrates. They never form scale, though very occasionally they may crystallise out in boilers if they are allowed to accumulate to an excessive degree by working the boilers for too long a period between cleaning times, or by failure to make proper use of the blow-off cock for keeping down the density of the water.

In land boilers the saline matter in solution should never be allowed to exceed 5,500 grains (12½ ounces) per gallon, which is equivalent to 2½ thirty-seconds when tested by means of a salinometer graduated in accordance with marine practice. It must be clearly understood that working to such a high concentration is not to be recommended, and should only be allowed in special cases where the feed water is of so saline a nature that serious loss of economy, or great inconvenience, would result had the boilers to be emptied frequently.

Generally speaking, the lower the maximum concentration the better. If boilers can be conveniently and economically emptied before 1,000 grains per gallon (half a thirty-second) has been exceeded, it will in most cases be wiser to adopt the practice. The higher the concentration in a boiler, the greater will be the liability to undue straining and other overheating troubles, and also to chemical reactions leading to corrosion, especially when nitrates are present in quantity.

Peat.—Though not a mineral constituent, peat can best be mentioned at this point. Many towns, now-a-days, obtain large supplies of surface water from mountainous districts. Such waters are usually very soft, and contain more or less peaty matter. Peat contains humic acids, or will produce them on decomposition. Waters of this class may, therefore, either be slightly acid, or be liable to become so on concentration, and, in the absence of more than very small quantities of calcium carbonate, may result in corrosion in boilers.

Contamination.—In addition to the natural mineral constituents of water there may be certain other constituents present due to pollution, some of which are of very great importance where the water is to be used in boilers. The principal of these are :—

Ammonium Sulphate, which is fairly frequently found in water, and which usually owes its presence to some contamination from gas purifying plants or coke ovens. It will not form deposit in boilers, but it has virulent corrosive properties due to the readiness with which it dissociates, when in contact with the hot plates of a boiler, with the liberation of sulphuric acid. Boiler plates have been known to be eaten through in a few months by this objectionable compound.

Sulphuric Acid may be run into rivers from the picking vats used in steel works, or from other works processes, and should it be present in such quantity as to render the water acid in the vicinity of the boiler feed suction pipe, very serious corrosion may result.

In the case of the discharge from steel works, the effluent would usually be a solution of sulphate of iron in acid. From explosive works a mixture of nitric and sulphuric acids may be run to waste in such a way as to lead to the pollution of a boiler feed supply.

Picric Acid.—During the war this chemical was found in certain canals and rivers due to effluents from explosive works. So far there has not been any great amount of experience to be gained as to its effect in boilers, but there is little room for doubt that it has decidedly corrosive tendencies and should, therefore, be avoided if such pollution should at any time occur. Picric acid is a powerful yellow dye and its presence in water is readily noticeable owing to its yellow colour, which is quite perceptible even when present in unimportant quantity.

CHAPTER III

CORROSION

The subject of corrosion is one that has exercised the minds of capable chemists and engineers for many years, some dealing with corrosion of metals in general, while others have confined themselves mainly to the corrosion of iron and steel. Innumerable experiments, many of them very elaborate and ingenious, have been carried out on a laboratory scale, under varying conditions, on different samples of metal, and with aqueous solutions of various salts or mixtures of salts, but unfortunately the results have not usually been very helpful in ascertaining the causes of, or discovering remedies for, the corrosion of boilers.

Generally speaking, it may be said that the only really useful information is that gained by noting the results in practice from the use of waters of known composition, and even then reliable information is difficult to obtain owing to the variations which occur from time to time in the composition of most water supplies, and the great differences in the conditions under which they are used. There is often a doubt whether the effects noted are the result of the water as analysed, or of some temporary changes in composition which may have occurred.

The question as to whether galvanic action is, or is not, the cause of wasting in boilers has often been the subject of discussion, and it is still more or less unsettled. In all probability both the advocates and the opponents of the galvanic theory are more or less right, and we come back to the old question as to which came first, the hen or the egg?

When the constituents of a water, or the products of decomposition, are such as favour corrosion, electric currents are set up in certain parts of the boiler owing to differences in potential, and the result is that those parts which are electro-positive to the others are likely to be attacked and corroded. If the composition of the water did not favour corrosion, then electric currents would not be produced, whereas if the metal of the boiler was of such a uniform character that no difference in potential existed there could be no electric current, and probably no corrosion would take place. It can hardly be disputed that galvanic action does have some influence, as we find so many instances of severe wasting at the junction of dissimilar metals, the metal which is electro-positive being the one attacked, and the wasting being local and close to the point of junction.

It is not even necessary to have a junction of two distinctly different metals in order to obtain differences in potential; two pieces of iron or steel of the same make, and of approximately the same composition, will almost invariably have a slight difference in potential, and even different parts of the same piece may show a difference in potential. Then again, if one part of a piece of metal is subjected to strain, such as that which results from bending, a potential difference is likely to be set up. This is probably a factor in the grooving of angles so frequently found in boilers. Again, commercial metals and alloys almost invariably consist of mixtures, more or less intimate, of two or more constituents of different potential, and thus are liable to set up numbers of minute galvanic cells. That this is a fact can be readily demonstrated by placing a piece of steel in a solution of common salt containing a few drops of solution of phenol phthalein and of potassium ferricyanide. Small points of colour soon appear, some blue and some red. The blue points prove that iron is entering into solution, while the red indicate the liberation of hydroxyl ions. Such a result can hardly be explained except by the presence of small galvanic cells, the points where iron is dissolved representing the positive plates, the negative plates being the points where hydroxyl ions are liberated. It will be seen, therefore, that, though it is claimed that steel can be commercially produced in America with only 0·05 per cent. of total impurities, it is commercially impossible to make boilers in which differences in potential do not exist to some extent, and it becomes necessary to endeavour to find what particular constituents in waters are liable to lead to corrosion, under what conditions they may be expected to do so, and how they can best be removed, or their corrosive tendencies neutralised. The working conditions likely to intensify corrosive tendencies must also be discovered.

Every one who has anything to do with the working of steam boilers is aware that an acid, such as sulphuric acid (vitriol), or hydrochloric acid (muriatic acid), obtaining access to the water will set up corrosion. The action of such acids is rapid, and occurs under any conditions, even if the water is cold. Fortunately they are readily neutralised by any alkali, such as soda ash, or even by calcium carbonate (limestone or chalk), and the resulting products are relatively harmless as far as corrosion is concerned.

Probably the natural mineral constituent in water which is best known for its corrosive tendency is magnesium chloride. It is harmless enough in itself, but is very liable to decomposition when concentrated, or in contact with hot plates, and under such conditions hydrochloric acid is liberated, and as this will occur in actual contact with the plates it readily attacks them.

Magnesium nitrate behaves similarly to the chloride, and is at least equally corrosive in its action.

Magnesium sulphate is frequently referred to as a non-corrosive salt. Probably it is so, or at least nearly so, if present alone, but generally a water will also contain an appreciable quantity of sodium chloride or nitrate. Under suitable boiler conditions these salts interact with the magnesium sulphate and probably in the first place magnesium chloride or nitrate is formed, which in turn is decomposed in contact with the hot plates, and thus causes corrosion. It is scarcely wise therefore to call magnesium sulphate non-corrosive, as it is apt to give a false idea of security.

Even magnesium carbonate must be looked upon with suspicion. It has been previously pointed out that it is seldom found in boiler scale, owing to the readiness with which it is converted into hydrate. Probably, at times at least, it is first converted into chloride or nitrate and thus results in wasting. Except on this assumption some cases of corrosion would be difficult to explain.

All magnesium salts are likely, therefore, to be objectionable, at any rate in waters containing appreciable proportions of chlorides or nitrates.

Any chlorides and nitrates may be considered as possible factors in corrosion, inasmuch as they cause otherwise non-corrosive magnesium salts to become corrosive, as mentioned above.

Nitrates, even in the absence of magnesium salts, appear, under suitable conditions, to have corrosive properties. This is usually most pronounced in high pressure boilers, and the higher the pressure the more likely they are to become active or more active. In practice their virulence is found to be largely dependent, not only on the temperature, but on the degree of concentration, and if magnesia is absent, or present only in very small proportion, it is possible, by keeping down the concentration sufficiently, to reduce the effect of nitrates to such an extent as to be unimportant.

Carbonate of soda has of recent years gained a reputation for being corrosive, but it is doubtful whether it can be proved to be so. Undoubtedly many waters containing bicarbonate of soda naturally, do cause wasting, but probably other constituents are responsible. It is not uncommon, for instance, for such waters to contain fair proportions of sodium chloride and magnesium carbonate, and it is far easier to understand how these can lead to corrosion than sodium carbonate. Nitrates may also be present, and the working conditions such as to render them active.

The sulphates of iron and aluminium may be considered as the most virulently corrosive salts naturally present in water. They readily dissociate, forming basic salts or hydrates, and with the liberation of acid, and where present in fair proportion their corrosive action is very rapid. The water in the gauge glasses is usually seen to become reddish, partly due to the iron first dissolved from the plates and afterwards precipitated, and, in the case of sulphate of iron, its own precipitation.

As previously mentioned, sulphate of iron (or even acids or nitre cake) may obtain access to rivers from steel works. At one time such effluents were knowingly run to waste into rivers, and even now-a-days, though it is not permitted, it does occasionally occur. In any case, leakages from works are liable to occur, and may do serious damage before being detected. Sometimes corrosive liquid may slowly ooze through the walls of a building built by the riverside, and it may even get into the river from beneath buildings erected on old tips in steel manufacturing districts. The author has come across several such cases. Another serious source of both iron and aluminium pollution is the slag heaps at collieries, which sometimes contain large quantities of pyrites, and yield such impurities to the rain water percolating through the heaps. In some cases, water draining from such heaps is of a dark brown colour due to dissolved iron, and also contains large quantities of aluminium and sometimes manganese. Where any of these sources of contamination are possible, the feed water should be carefully watched, and if any pollution is detected the source should be carefully looked for, and steps taken to prevent the pollution. If it cannot be arrested, a fresh feed water should be obtained if at all possible, or, failing this, carefully controlled treatment at once adopted.

Corrosion due to sulphate of alumina may also be found where oil is removed from the feed water by means of alumino ferric (commercial sulphate of alumina). In such plants it is most important that the alumina should be very efficiently removed in the treatment.

We now come to the ammonium sulphate or other ammonium salts found in the various effluents from gas plants, etc. The effects of these are similar to those of the iron and aluminium salts, ammonium sulphate decomposing in boilers with the liberation of sulphuric acid. The danger of ammonium salts is frequently unknown, or too little appreciated. Probably some confusion arises by considering ammonium salts as ammonia itself, which is well known to neutralise acids. If an effluent containing sulphate of iron or alumina had to pass through earth before entering a water supply, it is very likely the injurious salts might be thereby more or less removed, but this would not be the case with ammonium salts, and there is thus great danger, in the vicinity of gas works, of harmful effluents obtaining access to feed water supplies. Among the cases of corrosion by sulphate of ammonia which have been investigated by the author, one or two may be mentioned. In one case drainage from a concreted coke quenching bench contained ammonium salts, and found its way through cracks in the concrete to the soil beneath, percolating perhaps forty yards through the earth to a feed water storage pond, and resulted in very serious corrosion to a large battery of boilers. In another case water used for the washing of old sacks was run to waste through a culvert originally used for blowing off boilers, the culvert discharging into a stream utilised for boiler feeding purposes. Some of the sacks had previously contained sulphate of ammonia, and the effect on the boilers fed from the stream very soon caused alarm.

It is not even necessary that the feed water should be polluted. One case, suggestive of ammonium sulphate pollution, was of a gas works boiler fed by injector with town's water. It seemed impossible that any contamination could occur, but it was found that dust on the top of the flues of the boilers, and on the overhead beams, contained large quantities of ammonium salts, and some of this dust obtained access to the boilers when opened up for cleaning purposes, and led to corrosion.

The peaty matter found in some surface waters must also be considered among the causes of corrosion. It decomposes with the formation of humic acids, which, though very weak acids, will attack iron in the absence of neutralising salts such as calcium carbonate, and, as previously mentioned, this is usually absent or present only in very small proportion in the class of water liable to contain peaty matter.

In addition to the mineral constituents in water, the dissolved gases must be considered, as all boiler feed waters, unless previously treated, contain more or less oxygen and carbonic acid dissolved from the atmosphere. There are differences of opinion as to whether oxygen is corrosive in the absence of carbonic acid, but all agree that in the presence of even traces of moisture and carbonic acid, oxygen will readily attack iron and steel.

Gases probably play a part in very many cases of corrosion, and the removal of dissolved oxygen by passing the water through vessels filled with iron turnings has been adopted by a number of steam users. This has been strongly emphasised recently by Paul Kestner,* who has designed an excellent scheme for the purpose.

The corrosion due to dissolved gases in feed water is most noticeable when it takes the form of more or less circular pittings. This is likely to occur on surfaces where there is sluggish or no circulation of the water, especially if the parts are more or less horizontal ; or where, for any other reason, gas bubbles may separate from the water and remain on the metal for an appreciable time undisturbed. A particular spot having once been attacked is roughened and thereby rendered more liable to form a lodgment for further gas bubbles, and eventually the part may be pitted completely through while the surrounding

* "Degassing and Purification of Boiler Feed Water," read before a joint meeting of the Institution of Mechanical Engineers with the Society of Chemical Industry, March, 1921.

metal is unaffected. Any pocket in a steam plant, where air expelled from the water may collect, is liable to be corroded, though in such cases the action is more or less general instead of in the small circular spots above referred to.

As carbonic acid is liberated when bicarbonates are heated, corrosion similar to that due to dissolved gases may sometimes be experienced from the use of waters containing temporary hardness or sodium bicarbonate, unless suitable treatment be adopted.

The extent to which dissolved gases play some part in the general corrosion of boilers is almost impossible to decide with any assurance, as this corrosion cannot be distinguished from that due to other causes.

In connection with dissolved gases we must also consider the acid gases (other than carbonic acid) present in the atmosphere of towns, and particularly in manufacturing districts. Where feed water containing little alkalinity is exposed to air in open tanks, it is always possible for acid gases to be taken up to such an extent as to neutralise the alkalinity and render the water acid. Rain water is specially liable to this form of contamination, and in manufacturing districts it is seldom wise to use rain water for boiler feeding purposes without either chemical treatment, or the admixture of other water containing temporary hardness, which will neutralise acids dissolved from the atmosphere.

Corrosion in boilers may occasionally result from the use of greasy feed water. If pure mineral oil is used in the engines no corrosion can result therefrom, but some oils contain an admixture of animal or vegetable oil (known by the general name of fixed oils). Fixed oils are compounds of fatty acids with glycerine, and under suitable conditions they are decomposed in the boilers with the liberation of the fatty acids. Though not nearly so corrosive as the mineral acids, fatty acids certainly have the power of attacking iron and steel, and fixed oils should therefore be excluded from boilers for this reason ; in addition there is the possibility of any oils causing overheating.

While considering the causes of corrosion it will be well to draw attention to the possibility of corrosive constituents obtaining access to feed supplies from agricultural land. This form of pollution is likely to be more common nowadays owing to the extensive use of artificial manures containing nitrates or ammonium sulphate, and sometimes sulphate of iron, and to the greater area of land now being put under cultivation. Water supplies, hitherto used for boiler feeding purposes for years without ill effects, may suddenly be found to cause active corrosion. Unfortunately it is often difficult to convince boiler users that a supply which has given satisfaction for years can be held responsible for corrosion, and they are more ready to blame the material of the boiler plates, especially if a new boiler has recently been installed.

Calcium carbonate, having the power of neutralising acids, has a tendency to reduce the corrosive activity of other salts, and if present in fair quantity may even completely prevent corrosion from small quantities of the harmful salts. This must be carefully remembered when water is subjected to softening treatment. Imperfect softening may result in largely removing calcium carbonate while leaving the corrosive salts, which are thus rendered more active. This has frequently been found to occur in practice, and often the softening plant has been unfairly blamed.

From a consideration of the foregoing remarks it will be seen that in order to neutralise, as far as possible, the corrosive tendency of a water, mineral acid, if present, should be neutralised; and magnesium, iron, aluminium, and both free and half-combined carbonic acid removed as efficiently as possible. All this can be done by softening treatment with lime and soda ash, except that perhaps half a grain of magnesia per gallon may be left in solution. It is also wise, where otherwise permissible, to use a small excess of chemicals in the treatment, so that the water will have a slight caustic alkalinity which will enable it to combine with any carbonic acid which may be taken up by the water from the atmosphere between the softening plant and the boiler. The same chemical treatment will also decompose ammonium salts, if present, with the liberation of *free ammonia*, which will not be harmful in the boilers.

Softening treatment should also effect the removal of traces of oil, thus avoiding the liberation of fatty acids in the boilers.

Unfortunately nitrates will still remain in the water after treatment, but even these will be rendered less active if magnesia has been reduced to the lowest limit; and if an excess of chemicals be used they will also have a beneficial effect in retarding the corrosive action of nitrates. By keeping boilers cleaner, and thus avoiding unnecessarily high temperature of the plates, softening treatment will materially reduce the liability of nitrates to be decomposed, and will in this way tend to retard corrosion.

In addition to treatment, great care must be taken to avoid undue accumulation of nitrates in high pressure boilers. It is not possible to lay down any hard and fast rules for their maximum permissible concentration, as so much depends on conditions such as the working pressure and the temperature of the plates. Where boilers are clean and lightly fired, nitrates will be much less liable to cause trouble than in dirty, heavily fired boilers.

A peculiar and interesting form of corrosion is that generally known as graphitic wasting, which usually occurs in cast-iron economiser pipes. It is most difficult to detect as the surface of the metal appears to be unchanged, even though the metal may have been corroded almost through. On careful examination the surface of the metal is found to be so soft that it can be cut with a knife, and the cut surface has the appearance of graphite. Analysis shows the graphitic substance to be composed mainly of magnetic oxide of iron, together with graphite, silica, etc. The action appears to be a removal of some of the iron and the oxidation of the remainder of the impurities. The extent of the wasting can best be judged by filing the ends of the pipes, or cutting a section, when the graphitic portion will be seen sharply defined. It usually occurs when very soft water is used, either naturally soft or condensate water, and is probably due to the dissolved gases. Treatment of the water with a little caustic soda to remove carbonic acid, and tannin to absorb oxygen, usually prevents it.

It should be mentioned here that it is easier to prevent corrosion than to arrest it after pittings have been formed in a boiler plate. Where active corrosion has occurred, it is of the greatest importance that every particle of oxide be removed when the boiler is being cleaned, pittings being carefully scraped out. Blisters found on boiler plates sometimes contain actively corrosive substances of an acid nature, and in any case the black oxide acts as an electro-negative element, and the iron in the vicinity is, therefore, corroded.

CHAPTER IV

SOFTENING

It is not so many years ago that steam users were in the habit of putting such things as a dead pig or a sack of potatoes into their boilers as a means of preventing the formation of hard scale, but though the practice may have been permissible then, it can scarcely be recommended in these days of higher pressures and harder firing conditions. Probably any one trying the experiment in a modern boiler would pay dearly for his experience.

Nowadays it is not only necessary to exercise care in selecting the best feed water available, but to take steps to treat it in such a way as to render it as suitable as possible for the required purpose. The treatment will naturally depend on various circumstances such as :—

(1) The character of the water.
(2) The type of boiler.
(3) The rate of firing.
(4) The length of time it is necessary the boilers should work between cleaning times.

The most common practice at the present time, where scale is formed or corrosion occurs in boilers, is to introduce some chemical or boiler composition directly into the boilers, or into the feed water before it enters the boilers. In either case the chemical reactions take place mainly in the boilers. Though this, as a rule, cannot be considered the best practice, in most cases it is better than no treatment at all, especially as regards the prevention of corrosion.

One thing should be strictly avoided—namely, the *indiscriminate* use of the quack remedies known as boiler compositions. Some of these consist of suitable chemicals for the treatment of certain waters, but are only satisfactory for those particular waters, and then only if used in the correct proportion as ascertained by analysis, and introduced in the right manner. Unfortunately, too many are not only useless, but do harm by increasing the amount of deposit formed in boilers, while others are even liable to cause corrosion. A very few are specially prepared by the makers to suit the particular water, a sample of which they have analysed, and instructions are given as to the quantity required, and how it should be introduced. To this class no exception can be taken except on the score of expense. The makers must necessarily cover themselves for the cost of the analysis by including for it in the price charged for the composition, and the analysis has therefore to be paid each time the composition is purchased.

By far the better plan for steam users needing a boiler composition, is to submit a sample for full mineral analysis to a chemist who has a thorough knowledge of this branch of his profession, who can advise generally as to the suitability of the water for use with or without treatment, state the nature and amount of chemicals, if any, required for the treatment, and whether softening is desirable, or essential.

For the guidance of those who have already had an analysis made of the water to be used, but have not the knowledge necessary to interpret the results, it may be stated that in most cases either soda ash, caustic soda, or a mixture of the two will be all that is required, though occasionally the addition of a little tannin may be useful, especially where graphitic corrosion of economiser pipes is taking place. A method of arriving at the correct mixture, and deciding the quantity required will be found on page 112, but for the present it will be sufficient to state that the carbonates do not require either form of soda; the sulphate, chloride, and nitrate of calcium need soda ash; while the corresponding salts of magnesium should have caustic soda. As the carbonates present will convert some of the caustic soda into carbonate, a quantity of the soda ash equivalent to the carbonates should be replaced by caustic soda. This in some cases will mean that no soda ash will be needed, and in that case caustic soda equivalent to the permanent hardness is all that is necessary.

Whatever the particular chemical introduced, it is important, if the best results are to be obtained, that it should be introduced continuously, and *pro rata* with the feed, so that every portion of the water receives its quota of chemical before entering the boiler. Probably the best method of attaining this is to have a small pump specially for the solution, and to have it driven from the feed pump, so that every stroke of the feed pump results in the delivery of a definite quantity of solution, and it should be so arranged that the chemical and water mix before entering the boiler.

An alternative method of introducing the chemical solution is to have a small branch pipe, with a regulating cock, on the suction pipe of the feed pump. This pipe is immersed into the solution, and the cock so adjusted that for every stroke of the feed pump the correct quantity of chemical is drawn in for the treatment of the quantity of water delivered.

A less accurate method, but one frequently sufficiently good for the purpose, is to make a solution of the quantity of chemical necessary for an average day's supply of feed water, and allow it to run from a small tank provided with a regulating cock, into the feed tank, at such a rate as to be used up in the course of the day. If the solution tank be fixed fairly high, and the regulating cock sufficiently low, the difference in head during the day will not be enough to affect materially the rate of flow. This method is only really good when the boilers are being fed more or less steadily throughout the day, and the daily evaporation is fairly uniform.

Though a boiler of the Lancashire type may possibly work fairly satisfactorily with a water having twenty degrees of hardness, provided chemical treatment be adopted to cause the scale-forming constituents to be deposited in a readily removable form, and to neutralise those salts liable to lead to corrosion, it will usually be an advantage to put down a water-softening plant where the water has anything over ten degrees of hardness, and it is very unwise to allow water of even ten degrees to enter boilers of the water-tube type, more especially those in which small diameter tubes are used, the walls of which, being thin, allow little margin for wasting.

Where the character of the water, type of boiler, and working conditions make it undesirable that the water should enter the boilers without first removing as much of the scale-forming matter as possible, it becomes necessary to treat the water in a softening plant, and these may be divided into three main classes, viz. :—

(1) Exhaust Heater-Softening Plants.

(2) Lime and Soda Plants.

(3) The Permutit Plant.

Though an elaborate description of the various makes of softening plant would be out of place in a book which aims at being concise, it will be useful, to those about to make a selection, to give descriptions of the general principles of the different types, and then a few particulars of the special and distinctive features of the plants of the leading makers.

With one exception, all the types of softening plant used in this country are continuous in their action. The exception is that invented by Archbutt and Deeley, and it will be as well to describe this first.

Archbutt-Deeley Plant.—This plant has one or more units, each consisting of a large tank which can be quickly filled with a definite quantity of the hard water. The requisite quantities of lime and soda ash are transferred to the reagent tank, which is placed on a working stage above the main tank. Water is added, and the reagent prepared by boiling for a few minutes by means of a steam jet. The reagent is then run into the main tank through a number of horizontal pipes "A" (Fig. 1) near the bottom of the tank. These pipes are perforated on the upper side so that the reagent may be well distributed throughout the hard water. The contents of the tank are then thoroughly mixed and agitated by air blown in by means of a steam injector through perforations on the under side of another series of horizontal pipes "B" (Fig. 1). These pipes are near the bottom of the tank, so that the air stirs up the mud lying on the bottom of the tank. This agitation of the coarse grains of old mud with the hard water and reagent not only hastens the chemical reactions, but assists in the speedy coagulation of the fine particles of freshly formed precipitate, and results in quick settlement. After agitation has been continued sufficiently long (perhaps ten minutes), the water is allowed to stand until the precipitate has settled out, and the clear water is then drawn off from the surface through a floating discharge pipe " C " (Fig. 1). After the tank has been emptied to a definite level, the outlet valve is closed, and the tank refilled with hard water and the operations repeated. The old mud is allowed to remain in the bottom of the tank, except that sufficient is discharged occasionally to prevent its undue accumulation.

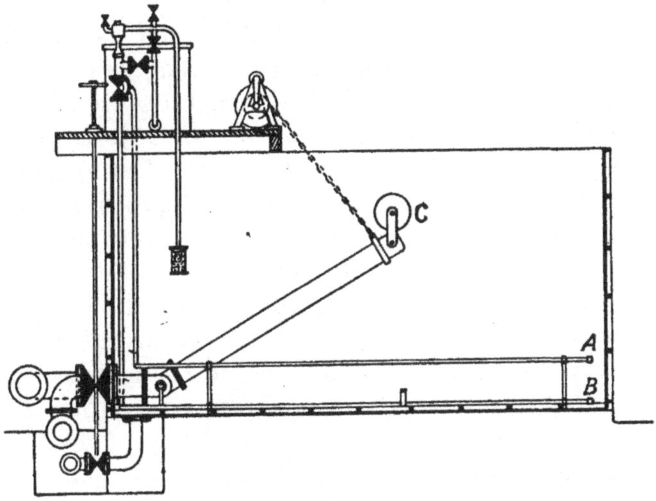

Fig. 1.—Archbutt-Deeley Plant.

If a continuous supply of water is required, two or more tanks are used, so that one or more are being filled, treated, or are settling, while one is being emptied.

It was found, in the early days of this plant, that objectionable deposits were sometimes formed in injectors, feed pipes, and clack valves, where the softened water became heated. This was proved to be due to the reaction between the lime and the last traces of magnesium carbonate not having been perfectly completed in the cold. The difficulty is overcome by passing the gases from a small coke stove into the discharge pipe while the water is being run off, and fitting baffles in the discharge pipe to ensure the admixture of the gas with the water. The carbonic acid thus dissolved by the water converts the excess of free lime into carbonate and prevents further reaction. Great care is necessary that undue carbonation be avoided, otherwise the excess of dissolved carbonic acid may lead to corrosion.

This type of softener is an excellent one, especially if the crude water is of variable composition, as each tank of water may be tested before being drawn off, and further chemicals added if necessary, or adjustments of the reagent made for the next tank. Unfortunately, it has not found the favour it deserves, probably owing to the fact that it requires more attention than continuous plants; in fact, for a plant of large capacity, one or more men may be required in constant attendance. Though this may be looked upon by many as a fault, it is really not so, as it tends to avoid the neglect which has resulted in many failures in continuous plants.

Continuous Plants.—In the continuous plants the chemical reagents, either separately or mixed, are automatically introduced in the correct proportions by means of apparatus controlled by the incoming water. The mixture of water and reagent then enters a reaction chamber in such a way as to keep the whole in a state of more or less agitation, mechanical means for better agitation being sometimes provided. After leaving the reaction chamber, the treated water passes somewhat slowly, and without agitation, through one or more settling compartments, where a portion of the precipitate is deposited, and then passes through a filter or filters, usually of carefully packed

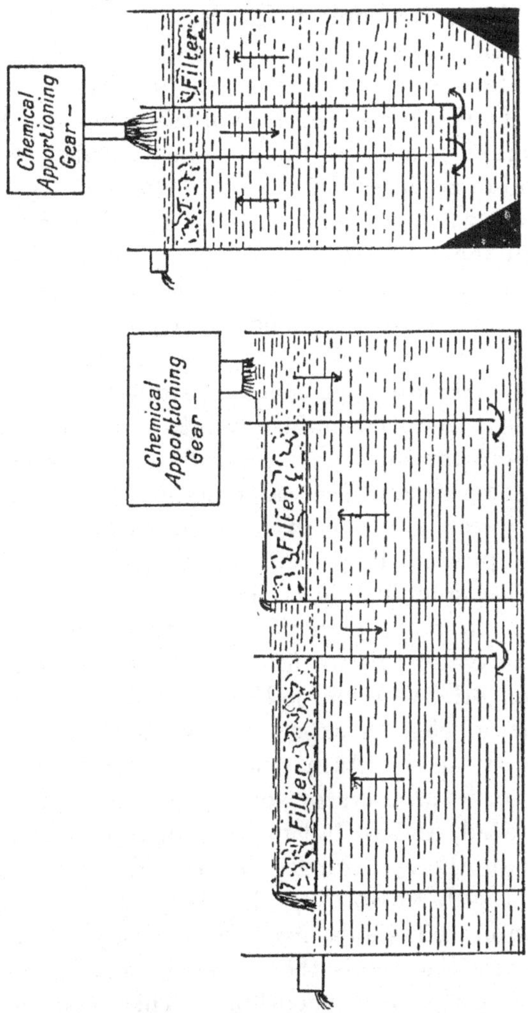

Vertical Plant.

Horizontal Plant

Fig. 2.—Typical Water-Softening Plant.

wood-fibre. Quartz sand filters are sometimes provided, either alone or following wood-fibre filters. Fig. 2 shows, diagrammatically, the usual arrangement of horizontal rectangular, or vertical softening plants.

Exhaust Heater-Softening Plants.—In this type the crude water enters a chamber in which it is brought into intimate contact with exhaust steam (or live steam if desired), by which the temperature of the water should be raised to boiling-point. If the heating chamber is well designed, and of sufficient size, most of the half-combined carbonic acid will be expelled from the *bi*carbonates, and result in the precipitation of the calcium carbonate with the exception of a few grains per gallon, and possibly some of the magnesium carbonate may also be thrown down.

If a water contains permanent hardness, a solution of soda ash is introduced uniformly into the water, either before it enters the heater, or immediately after leaving it. By this treatment the salts constituting the permanent hardness are converted into carbonates, and will be precipitated if present in quantity exceeding their maximum solubility in boiling water.

The heater in these plants should be so designed as to expose the water in thin films or drops to the action of the steam, otherwise the half-combined carbonic acid may not be efficiently expelled, and the best results will not be attained. As a rule the water is made to run over a series of trays or some similar device, sometimes perforated, and to fall in thin films or drops from one tray to another. After leaving the heater the water passes into a suitable settling chamber, where a portion of the precipitated solids may be deposited, and the partially clarified water is then filtered, usually through beds of carefully packed wood-fibre. This type of plant is very useful for utilising exhaust steam which would otherwise be blown to waste, as it not only saves the valuable heat of the steam, but results in a supply of hot feed water so much appreciated by engineers. It has the advantage of simplicity of control, only one chemical being required, and that a freely soluble one which gives a true solution easy of manipulation.

The oil present in the exhaust steam becomes entangled in the precipitate, and provided the water is hard enough to give a sufficiently large precipitate for the purpose, the oil may thus be efficiently removed. In order to render the removal of grease as perfect as possible, it is usual first to pass the steam through a mechanical oil separator, and the best plants are usually provided with such an apparatus.

Most of the makers of softening plants make plants of this type, or include, where desirable, exhaust steam heaters in conjunction with lime-soda plants, and, so long as the water is exposed in films or drops for a sufficient length of time to the action of the steam, the differences in the methods of doing so are of insufficient importance to justify special description. The Erith Engineering Company make only this type of plant, and they are, therefore, mentioned at this point.

Unfortunately this system does not usually result in very thorough softening unless lime is introduced as well as soda ash, owing to the solubility of magnesium carbonate, and in some cases, even though the plant be worked at its highest state of efficiency, the treated water, though much improved, may still have a fairly high degree of hardness. Again, the remaining hardness in such cases is likely to be due mainly to magnesium carbonate, which is liable to form a light floury deposit which may lead to priming and overheating troubles unless special care be exercised in the working of the boilers. The author remembers one case where, in spite of efficient heat treatment, the hardness was only reduced to 20 degrees, of which 18 degrees were due to magnesium carbonate.

Lime and Soda Plants.—These constitute the great majority of the softening plants used at the present time, and may well be subdivided into two classes, viz. :—

(1) Those in which the lime is introduced in the form of clear lime water.

(2) Those where the lime is introduced in the form of milk of lime.

Few of the former type are made now, though at least one prominent maker still adheres to the principle, claiming that where magnesium is present in large proportion it can be more effectively removed by lime water than by the more concentrated milk of lime. It is doubtful whether this claim can be upheld, as very efficient softening is frequently attained with the latter form, even with waters of very high magnesia content. Probably either form of introducing lime gives equally good results, but owing to the usually greater cost of plants for the lime-water process (on account of the need for a large vessel for the production of the necessary solution), the milk of lime plants find a readier market, especially as they are also more compact.

Lime-Water Plants.—Where lime water is used, a definite proportion of the crude water is by-passed to the bottom of a tall inverted conical vessel, known as a lime saturator, which is regularly charged with fresh supplies of slaked lime. In the lower portion of the cone the incoming water keeps the solid particles of slaked lime in a state of agitation, and the water thus becomes charged to its saturation point with lime.

As the solution rises in the cone, its rate of flow decreases as the diameter increases, and most of the solid particles are left behind. By the time the water reaches the outlet at the top of the cone it is in the form of fairly clear, saturated lime water.

In the plants made by Messrs. Royle, Ltd., the fraction of water necessary to be passed through the lime saturator is diverted from the main supply by allowing the crude water to enter a tank, known as a dividing box, which is fitted with weirs of adjustable width, so that any required proportion of the whole may be by-passed.

Mr. T. Roland Wollaston, when supplying plants of the lime-water type, divides the water necessary for the purpose by means of an adjustable partition in the tank into which the tipping bucket (described later) discharges, so that the desired fraction can be diverted to the saturator.

The Porter Water Softening and Engineering Co., whilst preferring lime water to milk of lime, recognise the disadvantage, from a selling point of view, of an expensive lime saturator. As an alternative they use milk of lime, as described later, but pass the mixture of water and reagent through a special mixing chamber before it enters the settling and filter tanks. The mixing chamber consists of a "V"-shaped tank with a vertical partition reaching nearly to the bottom (Fig. 3). The water and reagent enter at the top of one-half of the chamber, pass downwards to the narrow opening at the bottom, and then rise on the other side of the partition. The solid particles are thus prevented from settling out until they have been in contact with the main bulk of water for a reasonable time, giving the lime an opportunity for dissolving and reacting.

Some of the other makers attain the same result by

means of reaction chambers fitted with stirrers, a good example being a recent very large plant by the Paterson Engineering Co., put down for the treatment of five million gallons of water per day, in which the water and reagent is divided into four portions, each of which passes upwards through two tanks, working in series, each fitted with a vertical paddle.

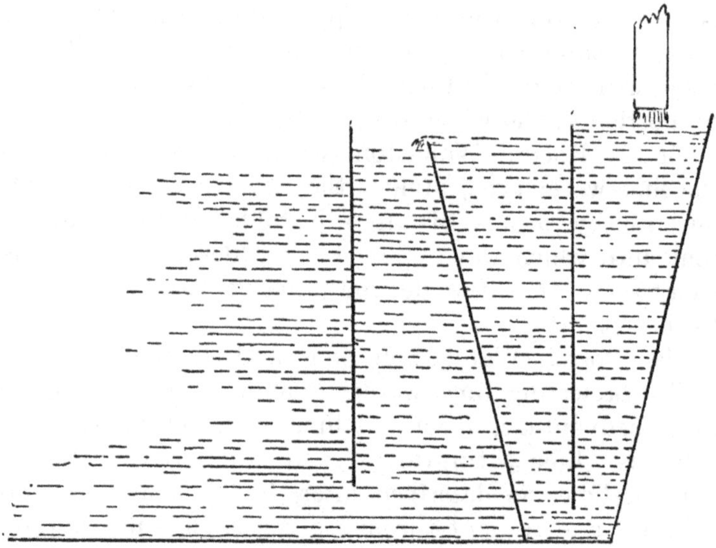

Fig. 3.—Porter Mixing Chamber.

With plants of the lime-water type, soda solution has to be made and introduced separately, involving further cost of manufacture in providing the additional gear for the purpose, and to a certain extent complicating the work of the plant. In the Wollaston plants, this is done by means of a scoop worked from the tipping bucket, exactly as for the apportioning of lime-soda reagent described later. Other methods are adopted by some makers, such as the displacement of a solution of soda by means of a small portion of the water by-passed for the purpose, or by passing a certain proportion of the hard water through special devices in which soda ash is dissolved, and from which a solution of definite strength is said to be supplied. A satisfactory description of these would occupy too much space, and those interested are referred to the catalogues of Messrs. Royle, The Underfeed Stoker Co., and Harris Anderson, Ltd.

In the opinion of the author, there is always apt to be a feeling of uncertainty whether such types are working correctly, and irregularities in the working cannot be detected by observation. In some cases, also, trouble is experienced in the constant strength solution devices, owing to soda crystallising out and choking up the apparatus.

Milk of Lime Plants.—In this type it is usual now-a-days to have both the lime and soda ash in one solution. The lime necessary for the treatment of a certain quantity of water is slaked, the requisite amount of soda ash added, and the whole diluted to a definite volume. This mixture, usually known as the *reagent*, is passed through a fine sieve to remove coarse grit and stones, into a container, known as the reagent tank.

Several methods are in common use for the apportioning of the reagent, the principal being :—

(1) Tipping-bucket.
(2) Lift pipe.
(3) Displacement.

The *tipping-bucket* method is the most generally adopted. It is very simple, and any irregularity in working is readily noticeable. There may be either a single or a double bucket, but in either case the crude water runs into a bucket which, when full, automatically tips and empties itself. The tipping movement actuates the gear for introducing a measured quantity of reagent sufficient for the treatment of the amount of water tipped.

The reagent measuring gear may consist of :—

(*a*) Cups or scoops which are filled by passing through the reagent when the bucket tips, the scoopful being delivered into the tipped quantity of water.

(*b*) Valves in the bottom of the reagent containers, which open when the bucket tips, and allow a certain quantity of reagent to pass. At one time these valves were frequently taper valves, which allowed the solution to flow during the time they were open, but as the quantity thus discharged varied with the head of solution in the tank, they are now more frequently made on a different principle, so as to deliver a measured quantity of reagent independent of the head.

It was found in practice that, for large plants, the great momentum of the tipping of a large volume of water resulted in such severe strains that the gear was apt to get out of order very quickly. Various methods have been adopted for overcoming this defect, as will be noticed in the following brief descriptions of the methods adopted by various makers for the apportioning of the reagent.

Most makers prefer the double-bucket type, which consists of a tank of triangular vertical section divided longitudinally to form two equal triangular tanks, as shown in Fig. 4. The tipper is supported, on a shaft about the point "A," inside a rectangular tank which receives the water when tipped. Stops are provided so that the tipper will come to rest when either of the two compartments is in a vertical position. The inlet water pipe is vertically over the centre of the tipping-bucket, so that it will discharge into whichever compartment is in the upright position. When the compartment is full the equilibrium is destroyed and the bucket tips, discharging its contents into the rectangular tank and bringing the other compartment into position to receive the hard water, and so the process goes on continuously. It will be evident that the accuracy of the apportioning of the reagent must depend on the sensitivity of the tipper. It must tip immediately a definite quantity of water has

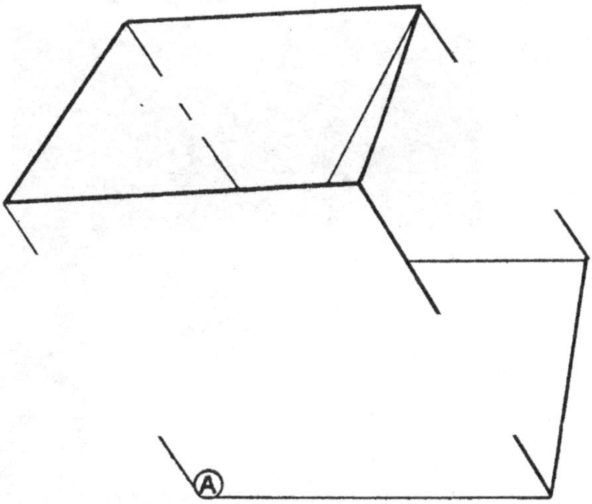

Fig. 4.—Double Tipping Bucket.

entered a bucket, but not before, and it must tip quickly at first, so that there shall be no delay in bringing the empty compartment into position to receive water. Tippers which are liable to discharge before the full quantity of water has entered, or which tend to lag when full, are untrustworthy, and will lead to irregular softening.

The Paterson Engineering Co., Ltd., have a tipper,

which they call an "Osilameter," provided with an extension lip, which is clearly shown in Fig. 5. After the water has reached the necessary level, a very small further quantity will flow into the extension lip, and exercise such an increased leverage that the equilibrium is immediately destroyed, and the bucket tips freely and quickly until the other compartment is beneath the water distributor. The upper part of the arm

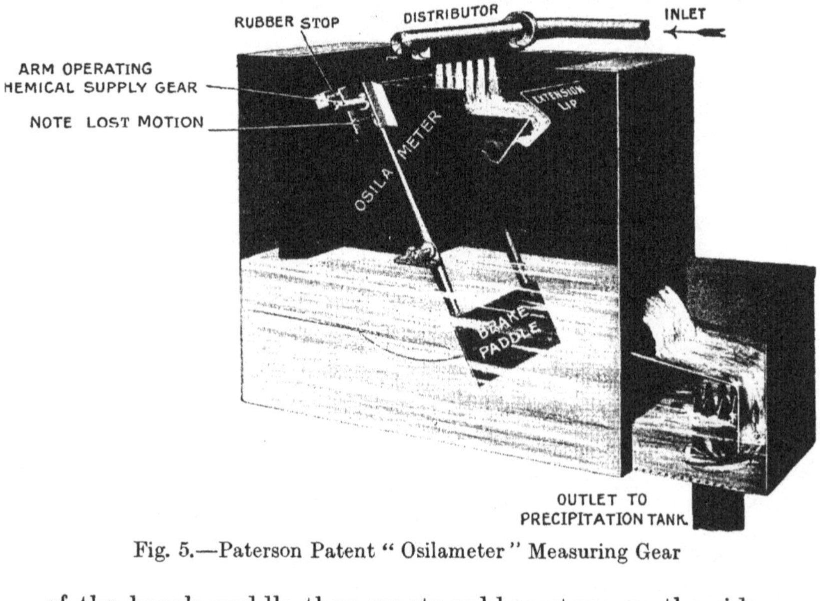

Fig. 5.—Paterson Patent "Osilameter" Measuring Gear

of the break paddle then meets rubber stops on the sides of the bucket, and thus brings the break paddle into action, slowing down the rate of movement by absorbing the momentum, and thus reducing the excessive strains which would otherwise result.

In the Paterson plants the reagent is measured by means of cups, which are filled by passing through the reagent, contained in a semi-circular tank, at each tip of the osilameter. The cups are loosely pivotted to the arms on which they are carried, so that, after rising above the surface of the reagent, they can tip and discharge their contents into a small trough, and thus to a mixing box where the reagent meets the measured water delivered from the tipper.

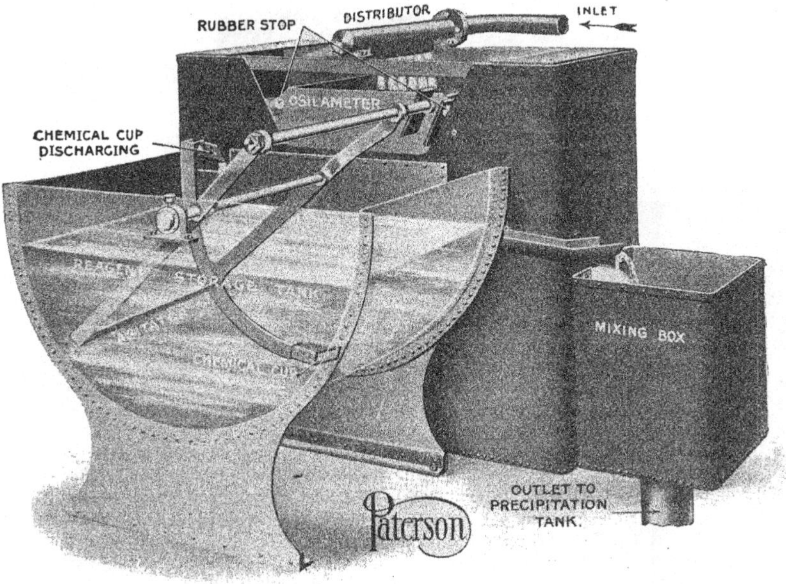

Fig. 6.—Paterson Patent "Osilameter" Chemical Supply Gear.

Fig. 6 illustrates this type of reagent measuring gear, and it will be seen that the reagent is kept stirred by an "Agitator," which passes through it once each tip.

For large capacities Paterson's make a more elaborate plant, illustrated in Fig. 7. The incoming water, while passing over a weir, is divided by an adjustable partition, only a small portion entering a relatively small osilameter, thus avoiding the great momentum which is unavoidable when large volumes are passed through tipping buckets. The larger part of the water actuates a water wheel, which continuously revolves a paddle wheel in the reagent tank, and thus ensures a perfect mixing of the reagent. On the same shaft as the osilameter is a smaller double tipper into which the reagent

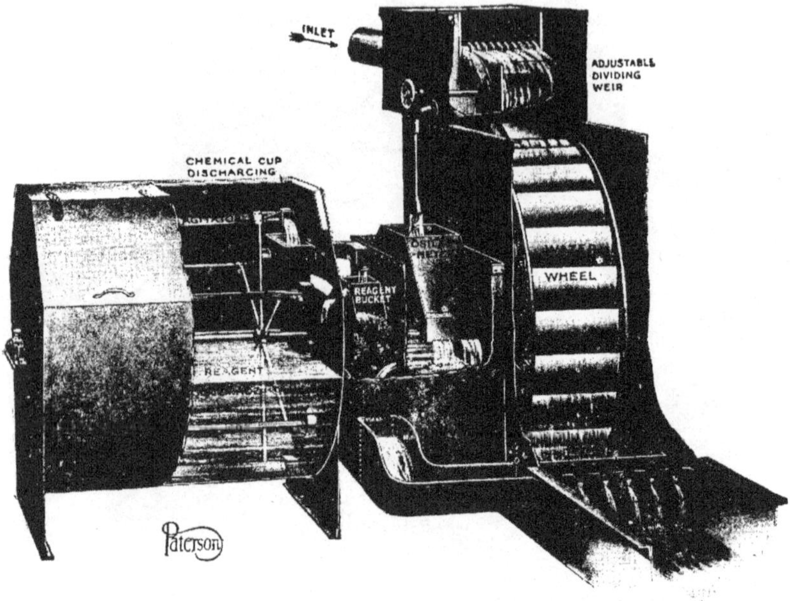

Fig. 7.—Paterson Patent "Osilameter" (By-Pass Type).

is continuously fed by cups fixed on the paddle wheel. The quantity of the reagent delivered to the small tipper is such that a compartment is always filled with reagent before the osilameter is ready to tip, and the excess reagent overflows and runs back to the reagent tank. The water and reagent delivered by the tipper run through a trough to meet the main stream of water, their preliminary admixture being effected by passing along a trough fitted with a notched weir and baffles.

Fig. 8.—"Lassen-Hjort" Patent Continuous Automatic Measuring Apparatus.

The United Water Softeners, Ltd., have tipping-buckets of the simple design first described, and although not provided with special means for increasing the sensitivity, they are fitted with a special device, controlled by ball floats, by means of which the bucket is locked until released by the lifting of the ball by the rising water "A" (Fig. 8). This method effectively prevents too early tipping, but does not increase the sensitivity. It is of the greatest importance that the locking gear be kept in good order, otherwise the ball float may not release the bucket, and the supply of reagent will be stopped, and hard water will enter the plant untreated.

In order to minimise the shock when the bucket tips, this firm have a special buffer device (see "B," Fig. 8), which consists of an air chamber beneath each compartment of the tipping bucket, the under side of these buffer chambers being open. When the bucket tips, it does so quite freely until the empty compartment is in position to receive the incoming water, by which time the open (under) side of the buffer chamber meets, and is sealed by the water in the rectangular tank in which the tipper works, and an air buffer is thus formed and the momentum absorbed. In order that the bucket shall not be prevented by the imprisoned air from falling to its proper position of rest, small adjustable air vents are fitted to the buffer chambers, by means of which the rate of fall can be nicely regulated.

The reagent measurement in the United Water Softeners' plants is affected by their "Positive Measurement Valve" ("C," Fig. 8, and Fig. 9). A spindle, which is raised and lowered again at each tip of the bucket, is fitted with two pistons, which work in a small cylinder at the bottom of the reagent tank. The distance between the pistons can be adjusted, but is such that the cylinder is always open at one end. When the tipper is at rest the upper end of the cylinder is closed and the lower end open. When the bucket tips, the spindle is first raised, thus opening the upper end of the cylinder which becomes filled with the reagent. The spindle then falls

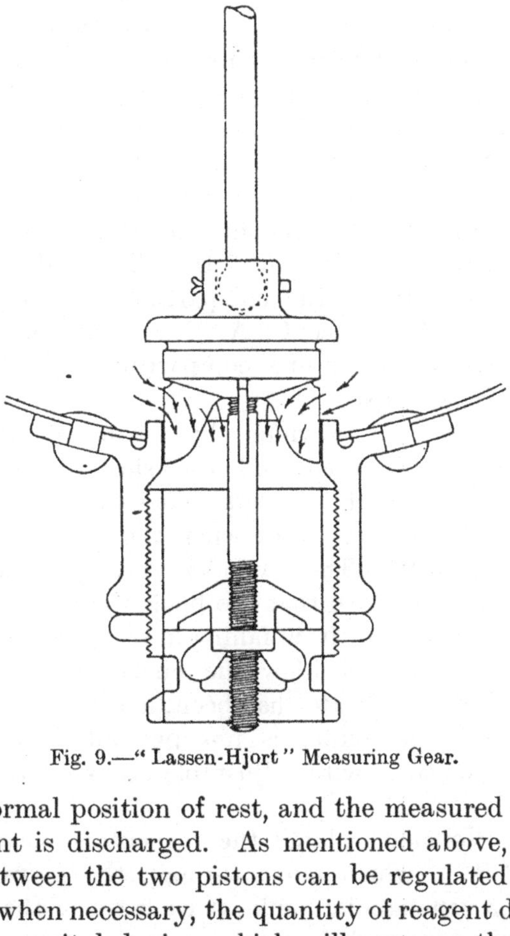

Fig. 9.—" Lassen-Hjort " Measuring Gear.

to its normal position of rest, and the measured quantity of reagent is discharged. As mentioned above, the distance between the two pistons can be regulated in order to vary, when necessary, the quantity of reagent delivered. This is a capital device, which will measure the reagent very accurately, but it is very important that it should be regularly cleaned, as otherwise the measurement cylinder will quickly become coated with reagent and its capacity reduced.

For large plants the United Water Softeners fit a regulating cock on the inlet water pipe, by means of which any desired proportion of the water may be diverted, and a smaller portion only used to actuate the tipping bucket, thus overcoming the difficulties which occur when large tipping buckets are used.

In the Wollaston plants, improved sensitivity of the tipping bucket is attained by having a space between the two compartments, and thus increasing the leverage. This is illustrated in Fig. 10, " D." The middle portion of the space between the compartments is bridged by an inverted " V " piece (" A "), so that the water is quickly diverted from one compartment. to the other when the bucket tips.

A buffer gear is provided, by means of which the momentum is absorbed and the shock reduced. It consists of two small cylinders, fixed to the top of the tank, in which pistons work when the bucket tips, and form air buffers. The cylinders are provided with adjustable air vents, so that the bucket is not prevented, by the imprisoned air, from falling into its proper position of rest. In order to allow the bucket to tip quickly and freely until the water has been diverted to the other compartment, the buffer is not put into action until the bucket engages with adjustable stops on the piston rod.

In the Wollaston plants the reagent is measured in scoops (" B "), which move through the reagent, contained in a semicircular tank, at each tip of the bucket. The scoops are carried on pipe arms (" C "), through which the contents of the cups are conveyed to a central trough when the cup is raised above the horizontal

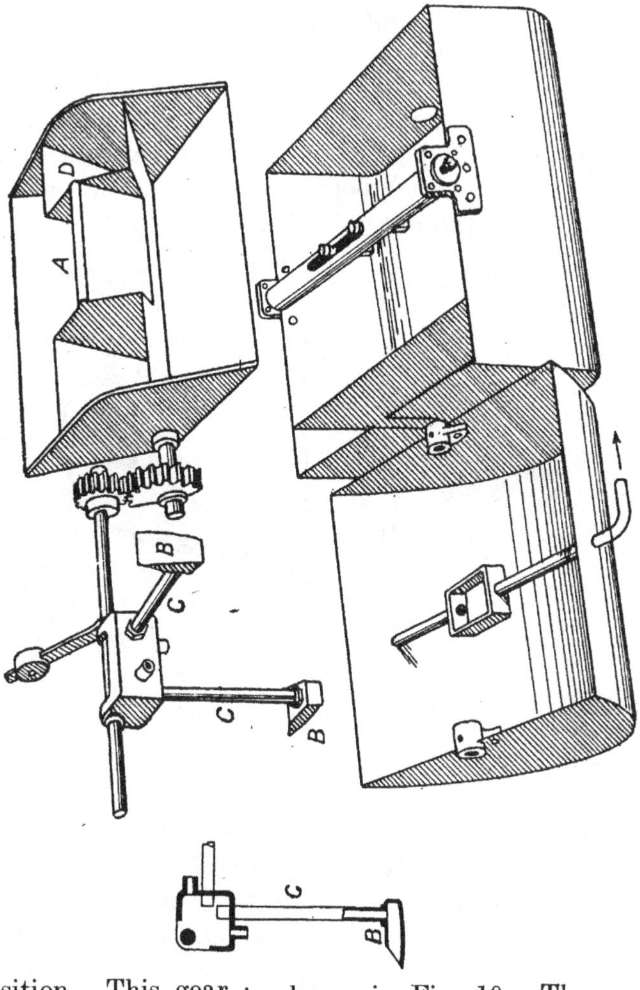

Fig. 10.—Wollaston Tipper and Gear.

position. This gear is shown in Fig. 10. The reagent is stirred by a paddle which passes through it at each tip of the bucket.

Messrs. William Boby & Son, and The Becco Engineering and Chemical Co., Ltd., each make the same type of plant, with a single-tipping bucket, illustrated in Fig. 11. When the bucket is filled to a certain level it tips its contents into the rectangular tank and then resumes its horizontal position. While the measured quantity of water is being emptied, a small compartment comes into position to temporarily receive the incoming water, this water running into the main bucket when

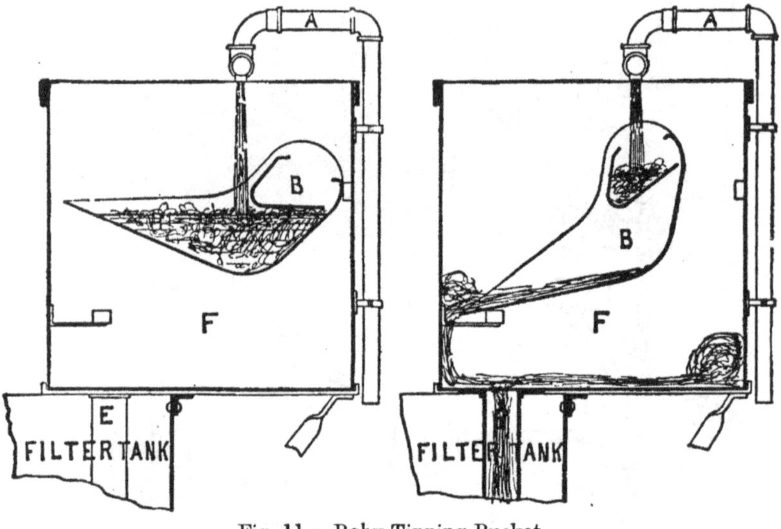

Fig. 11.—Boby Tipping Bucket.

the normal position has been regained. The motion of this type being a partial rotation on an axis above the centre of the bucket, is much more gentle than that of the double-tipping buckets, and causes very little shock.

The reagent measurement in the Boby and Becco plants is effected by means of a single scoop, illustrated in Fig. 12, which dips into the reagent while the bucket tips, rising and delivering its contents direct *into the bucket* as it resumes its horizontal position. The capacity of the reagent scoop can be varied by means of the adjustable central overflow pipe shown in the illustration. These scoops are of decidedly larger capacity than the reagent cups previously described, and this is an advantage, as a more dilute reagent is used, and consequently the measurement is likely to be rather more uniform.

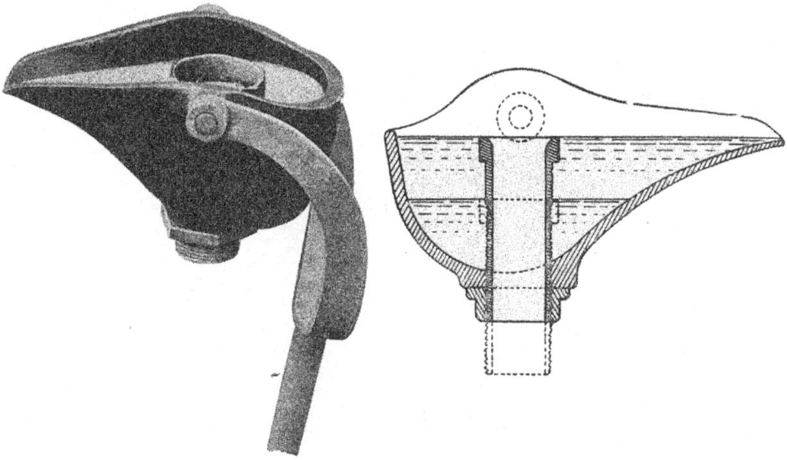

Elevation. *Section.*

Fig. 12.—Boby Measuring Scoop.

Another slight advantage in these plants is that the paddle which stirs the reagent passes twice through the reagent at each tip of the bucket.

Messrs. The Kennicott Water Softener Co. make an entirely different type of plant. The hard water first passes over a water wheel, which actuates stirring gear in the reagent tank. It is then divided by passing over adjustable weirs, and a very small proportion passed into a small tank, known as the regulating tank, which contains a float (see Fig. 13). The reagent tank is *much* larger than in the plants previously described, as it is essential, in this system, that the reagent should be very dilute, and consequently more fluid. The reagent tank contains a pipe (the *lift pipe*), the lower end of which is connected by a flexible joint to the outlet pipe at the bottom of the tank. The other end of the lift pipe is connected, by a chain passing over pulleys,

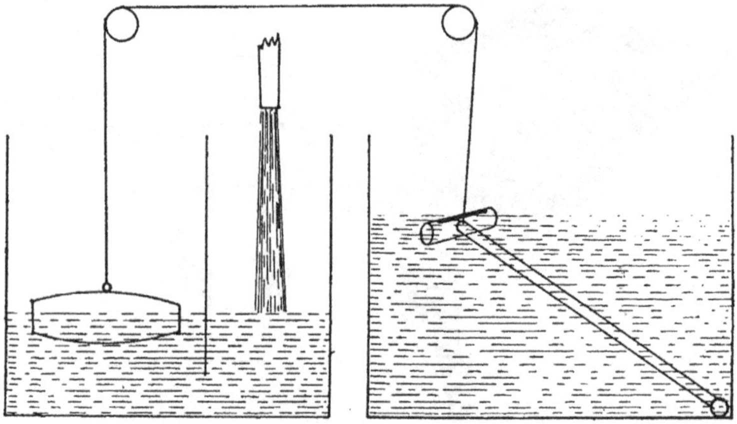

Fig. 13,—Kennicott Gear.

to the float in the regulating tank, the length of the chain being such that when the float is at the bottom of the regulating tank, the upper (open) end of the lift pipe is just above the surface of the reagent. As the small definite proportion of the incoming water enters the regulating tank the float rises, and the lift pipe is correspondingly lowered into the reagent, which is thus allowed to flow, *pro rata*, with the incoming hard water, into the main stream of water at a suitable point. By the time the regulating tank is filled, the reagent tank is emptied, and the quantity of water by-passed to the regulating tank is so adjusted that this results when the quantity of hard water for which the charge of the reagent is intended has entered the plant.

This method, as claimed by the makers, avoids the use of cups or valves for the measurement of the reagent, but as the reagent has to run through a comparatively long and narrow pipe, which will undoubtedly tend to choke, it is doubtful if anything is gained in that way, especially as it is easier to clean cups and valves than the lift pipe.

The plant is undoubtedly a good one, but has the slight disadvantage that it is necessary to stop the treatment for a short time whilst the reagent tank is being filled and the regulating tank emptied. On the other hand, the fact that there are no tipping or oscillating parts is a distinct advantage.

The Underfeed Stoker Co., Ltd., have introduced another entirely different system, which is decidedly ingenious, though whether it is found to work satisfactorily in practice the author is unable to say. The main features are shown diagrammatically in Fig. 14. The incoming water flows over a circular weir (A), a small proportion being divided off and carried by a pipe to a special dividing box (B). The main bulk of water then actuates a water-wheel by means of which a paddle is revolved in the reagent tank. The dividing box (B) has two weirs, one of which is of fixed width, for the purpose of further reducing the proportion of water required for the apportioning gear. The other weir (C) is provided with a sliding division automatically moved by means of a carefully designed worm-wheel (D) slowly rotated by a cord attached to a float in the float tank (E). One portion of the water flowing over the weir (C) passes to the float tank, whilst the other portion flows into the reagent tank and displaces an equal volume of reagent, which then mixes with the main stream of water. It will be obvious that the reagent will gradually become more dilute, but this is compensated for by the regulated widening of the adjustable weir by the worm-wheel, the rate of widening becoming greater as the strength of the reagent becomes less and less.

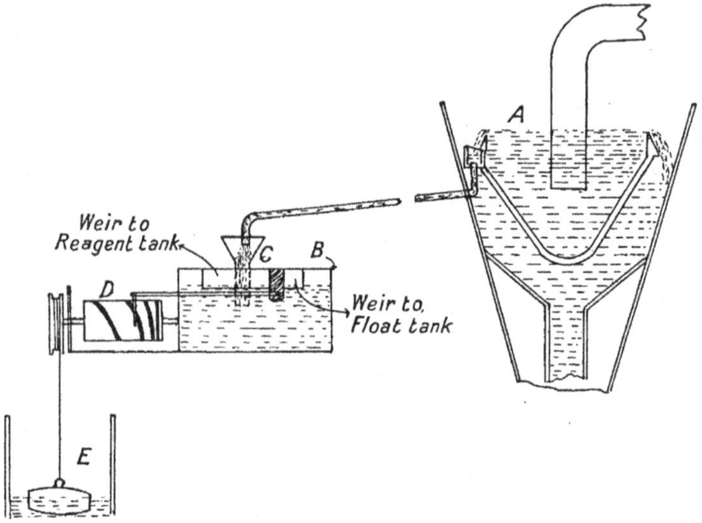

Fig. 14.—Underfeed Stoker Co.'s Gear.

The Porter Water Softening Co. also avoid the use of bearings in contact with the reagents. In their plants the crude water first enters a tank containing a float, the rise and fall of which lowers or raises the lip of a return overflow weir ("A," Fig. 15) in a small compartment which is kept constantly supplied with reagent by means of cups on the arms of a paddle wheel in the reagent tank, similar to that of the Paterson Engineering Co. (see Fig. 7). The depth of reagent in the compartment is thus varied *pro rata* with that of the crude water in the float tank, and the excess

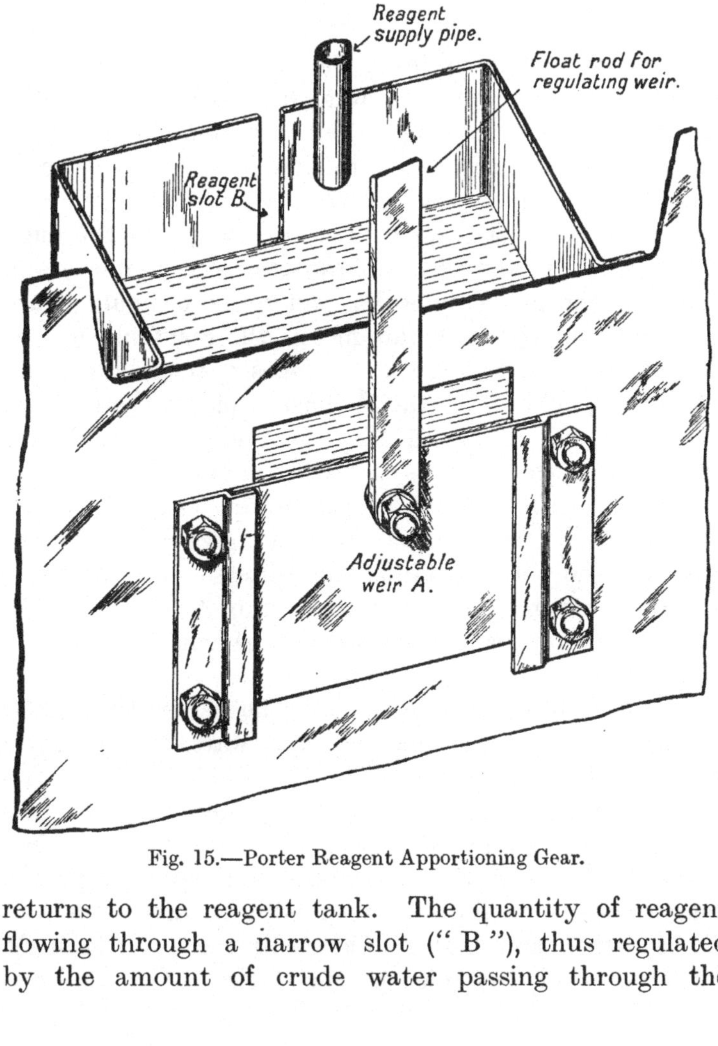

Fig. 15.—Porter Reagent Apportioning Gear.

returns to the reagent tank. The quantity of reagent flowing through a narrow slot (" B "), thus regulated by the amount of crude water passing through the plant, runs down a chute to mix with the crude water as it enters the special mixing chamber previously described (Fig. 3). The weir (" B ") is kept clear by means of a scraper constantly moving backwards and forwards across it.

As previously mentioned, the water, having received its quota of reagent passes first into a reaction chamber, and then into a settling chamber or chambers. It is hardly necessary to enter into details of these, as most makers vary them to suit the requirements of each particular case. It is important, however, that thorough mixing should take place in the reaction chamber, and also that the incoming water and reagent should churn up the old mud, as by this means the reactions are expedited, and the precipitate becomes coarser, and settles out more quickly. In the Wollaston plants the reaction chamber is so designed that the lighter particles of precipitate, which, it is claimed, contain all the oil, rise to the surface, and as the water leaves from the bottom of the reaction chamber no grease can go forward to the settling tanks and filters. The principle is certainly a good one, but the author has not had the opportunity of personally ascertaining whether absolute removal of grease is thus assured.

The settling compartments may be either rectangular or circular, but in either case the water should enter at the bottom and leave from the top of each compartment, and no plant should be accepted in which the water passes downward through the settling chambers. If the plant is of the horizontal rectangular type the compartments should be correspondingly small, otherwise much of their area may be wasted by the water taking a short cut from the inlet to the outlet. This result is minimised by subdividing the main tank into two or more small compartments, as shown in Fig. 2. The author favours vertical cylindrical plants where the circumstances render them otherwise convenient, but there are differences of opinion on this point. Generally, in the vertical type, the mixture of water and reagent passes down a central pipe to the bottom of the cylindrical tank, mixes with the old precipitate, and rises fairly uniformly throughout the whole section of the cylinder with no possibility of taking short cuts.

In order that the precipitate may be removed as efficiently as possible by sedimentation, the rate of flow of the water should be slow, and the tanks should be of ample size. This will be again referred to in Chapter V.

Though, by efficient sedimentation, most of the precipitate may be removed, there will still be traces remaining which should be removed before the water leaves the plant. Most makers effect this by means of wood fibre filters. These consist of long narrow shavings of wood, packed firmly and uniformly into suitable compartments, the water passing through in an *upward* direction. Wood fibre cannot be successfully used for downward filtration.

Wood fibre filters are quite good enough for most softening plants, provided that efficient sedimentation has first been effected, and that sufficiently frequent cleaning or changing is maintained. If the cleaning be neglected water can pass through wood fibre without the precipitate being removed.

When wood fibre filters require clear ng, the old material should be removed and replaced by n w or cleansed fibre. The wood fibre which has been remo ed can be cleansed by thorough washing, or it may be llowed to dry and then be well shaken.

The United Water Softeners, Ltd., have introduced another method of cleansing wood fibre filters while in

Fig 16.—"Lassen-Hjort" Patent Filter Cleaning Attachment.

position and without stopping the plant. The filter bed in rectangular plants is divided, by means of a longitudinal partition, into two sections. When cleansing is desirable, the sediment in one section is agitated by means of air blown in through perforated pipes below the filter bed. The sediment is thus released and carried upwards through the fibre with the water, and the muddy water is allowed to run to waste through a central trough, which can be partly rotated so as to collect the water from whichever section is being cleansed. Fig. 16 shows the working of this device. Though such an arrangement may do all that is claimed for it, the author considers it better to avoid the necessity for it by making the plant of ample capacity to ensure efficient sedimentation, and thus enable the plant to work for several months without cleaning the filters.

The Wollaston plants avoid the use of wood fibre by means of sectional coconut matting filters. The filtering area of the plant is divided into small sections, on each of which is placed a light iron frame, over which is sewn a piece of coconut matting. When the filter requires renewal, each of the small units can be separately lifted out and replaced by a spare. The fouled filters, after being allowed to dry, are beaten to remove the sediment, and they are then ready to be used again when required. This type of filter is illustrated in Fig. 17.

Though, in nearly all cases, wood fibre or coconut matting filters are quite satisfactory in a well designed plant for boiler feed water purposes, it may occasionally be necessary, especially where freedom from oil is essential, to instal a quartz sand filter to remove any traces of sediment which might pass the preliminary filters. They are so seldom required that it is unnecessary to describe the various designs in detail. Filtration through sand is always carried out in a *downward direction*, and may be effected by simply allowing the water to flow by gravity through a bed of quartz sand supported on pebbles, or it may be forced through such a filter under pressure.

The chief differences between the sand filters of various makers are in the methods of cleansing the sand, which gradually becomes choked and prevents the passage of the water. The main principle is to agitate the sand bed and reverse the direction of the flow of water, allowing

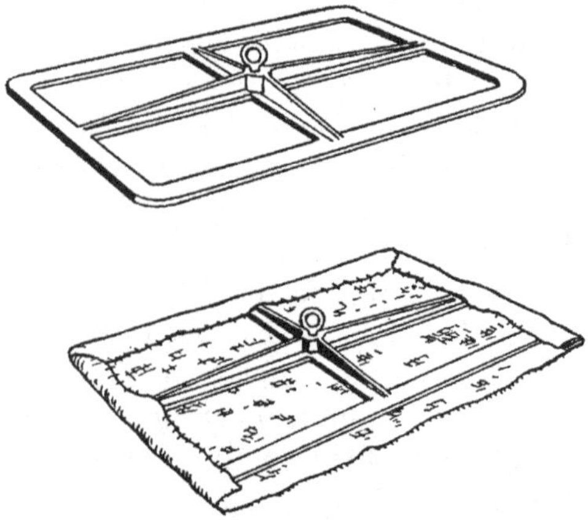

Fig. 17.—Wollaston Coconut-Mat Filter.

the dirty water to run to waste. A simple means of agitating the sand, which is frequently adopted, is to have one or more stirrers in the sand bed, which can be rotated either by hand or by power while the water is reversed. The chief fault of this method is that there is little disturbance of the bed near the axis of the stirrers. Paterson's have a special device, blowing in air through perforated nozzles at the bottom of the filter bed and effectively churning up the whole mass. The Turnover

Filter Co. have a filter bed in the lower part of a horizontal cylinder. The cylinder can be rotated on a horizontal axis, the sand bed being disturbed by rocking the cylinder backwards and forwards through a suitable angle.

Sludge cocks for periodically removing the accumulated precipitate are provided in all types of plant. Many plants, particularly of the vertical type, have hopper bottoms to enable the sludge to flow to the sludge cock, and these are really useful. Some makers have adopted special means for effecting sludge removal, such as rotating scrapers to propel the sludge towards the outlet, sludge agitators to render it more fluid, and perforated rotating pipes to pick up sludge from all parts of the bottom of the tank. All these may be effective for the purpose intended, but scarcely necessary, and, in the case of vertical plants and reaction tanks, it is undesirable to keep the tank bottoms too clean. Such devices need not, therefore, influence the selection of a softening plant. As previously mentioned, old precipitate assists in coarsening the new precipitate and thus hastens sedimentation, and it is desirable, therefore, *not to remove the mud entirely*, but only to prevent its excessive accumulation.

A useful refinement for softening plants is to have a reagent mixing tank at ground level, and to pump the mixture to the reagent tank on the top of the plant. All makers will make such apparatus if desired.

There is always a possibility that the chemical apportioning gear of a softening plant may receive less supervision than is desirable, owing to the fact that it is almost invariably on the top of the plant, and usually, therefore, necessitates the mounting of a ladder to reach it. Where possible, the plant should be so situated that the top of the plant is about on a level with some portion of the ground, or with one storey of a building, from which it can be conveniently reached by means of a gantry. The author was connected with the installation of one very large plant, made by the United Softeners, Ltd., where the height of the plant would assuredly have resulted in more or less neglect had the gear been placed on top. This difficulty was satisfactorily overcome by placing all the measurement apparatus at ground level, the water and reagent being separately pumped to the top of the reaction and settling tank. The working parts were, therefore, under constant observation, and the plant worked very satisfactorily.

Permutit Plant.—This is a system of softening in which no chemicals whatever are added, and normal variations in the composition of the water do not affect the results of treatment, which is therefore rendered very simple.

The water to be treated is passed through a bed of an artificial mineral known as Permutit. This consists of a complex silicate of aluminium, sodium, potassium, calcium, magnesium, etc., and is an imitation of a natural mineral known as Zeolite.

When water containing hardness-forming salts comes into contact with Permutit, a chemical interchange takes place, the calcium and magnesium being taken from their salts, and sodium and potassium substituted. The hardness of the water may thus be entirely removed.

After Permutit has removed a certain amount of hardness, its power is reduced and eventually exhausted, but it is revivified by passing through it a fairly strong solution of common salt (sodium chloride), which brings about a reverse reaction, removing from the exhausted Permutit the calcium and magnesium taken up from the water, and replacing them by sodium, thus restoring the original condition, and enabling it to soften water once more.

Usually, in practice, plants are made of such size as to efficiently treat the particular water at the necessary rate for a certain number of hours, after which salt solution is passed through the plant to regenerate the Permutit. It is claimed that this can be continued indefinitely. It was originally arranged for the time for regeneration to be equal to that for softening, but it is said that new developments promise increase in efficiency and running economy.

For waters very low in natural hardness, or practically free from temporary hardness, this process is a very useful one, and in any case it is so for the treatment of water for laundry purposes and processes where entire freedom from hardness is desirable. When temporary hardness is present, however, sodium bicarbonate is left by the treatment, practically equal in amount to the temporary hardness removed. As it is usually considered objectionable to allow sodium carbonate to accumulate in boilers, this must be looked upon as a distinct drawback to the process, especially as the cost of removing temporary hardness by Permutit is exactly the same as for permanent hardness, whereas by lime treatment temporary hardness can be removed, on an average, at about one-sixth the cost of permanent hardness.

Where absolute freedom from hardness is desirable and large excess of sodium carbonate would be objectionable, such as in locomotives or similar types of boiler, the Permutit system can be used with great advantage for the final treatment of water which has been softened by lime, or even lime-soda treatment. The bicarbonates having been removed, no additional sodium carbonate is formed by the Permutit. Unfortunately, the cost of the double plant prevents the general adoption of this method.

It is perhaps only fair to say that the makers of the Permutit plant maintain that the carbonate of soda produced by the process is not objectionable in boilers, provided that the internal gun-metal fittings are of good quality. It is difficult to believe that sodium carbonate thus formed can differ in its effects from that added as soda ash in excess of what is required for the precipitation of permanent hardness, or from that present naturally in certain waters.

CHAPTER V

SELECTION OF SOFTENING PLANTS

WHENEVER possible, unbiased expert advice should be obtained by those considering the purchase of a softening plant, as it is of the greatest importance that first cost should not unduly influence the choice. An efficient plant, which will continue to perform its duties for a reasonable time without repair or frequent adjustment, is a blessing to a steam user. On the other hand, an inferior article may cause endless trouble, and many plants have been discarded owing to the difficulty of obtaining satisfactory results. Imperfect softening, instead of removing the corrosive tendency of a water, sometimes increases it, while in other cases it may lead to choking up of pipes and valves.

A few words of advice based on years of experience will not be out of place.

Size.—In the first place the plant should be of ample size. This does not mean that a plant should be obtained with a rated capacity higher than is required. The chemical apportioning gear should be designed for satisfactory working at the maximum rate necessary, but no more. It is the reaction and settling chambers which need to be large, and they should always be of such size that the water will take *at least* two hours to pass through them—*i.e.*, a plant for the treatment of one thousand gallons per hour should have reaction and settling tanks with a capacity of not less than two thousand, and preferably three thousand gallons. In special cases even larger tanks are an advantage.

Strength.—All parts of the plant should be of ample strength for standing the necessary strains or pressure. This applies particularly to exhaust steam heaters, which, having to stand sudden changes of pressure, and a wide variation of temperature, are liable to be distorted or even ruined unless substantially made. They should be constructed of stout materials, and be well stayed.

Reagent Gear.—Moving parts of the chemical apportioning gear should be well designed and strongly made to withstand reasonable wear and tear for long periods without undue attention, adjustment, or repair. It should be so designed as to work uniformly whether the reagent tank is full or nearly empty, and all parts liable to choking up should be visible and readily accessible for cleaning purposes. In the case of tipping-buckets, they should be so sensitive in action that they can be relied upon always to tip immediately when filled to a definite point, not being unduly influenced by rust or dirt in the bearings, or deficient lubrication. Any special arrangements for increasing the sensitiveness should receive careful consideration.

The apportioning gear should work satisfactorily, when required to do so, at much below the rated capacity. Special devices for keeping the reagent continually stirred must therefore receive consideration even though the price of the plant may be thereby increased. When the plant is working slowly the suspended particles of lime in the reagent are liable to be deposited on the bottom of the reagent tank unless constantly stirred, thus leading to want of uniformity in the strength of the reagent.

Movable joints which come in contact with reagent are always liable to give trouble through sticking unless carefully attended to, and though many excellent plants have such joints, it must be considered as a favourable point in a plant that avoids them.

Recharging.—If inconvenience would result from the stoppage of the plant for the purpose of recharging with reagents, care should be taken to select a plant which can be recharged while working.

Reagent Capacity.—The reagent tank should be of such size as to render frequent charging, or recharging at inconvenient times, unnecessary.

Preparation of Reagent.—Convenient arrangements for the preparation of fresh supplies of reagent in readiness for recharging should be provided. These should be readily accessible, so that the man in charge has not to climb ladders, or be in awkward positions while mixing the reagent. This is particularly important in bad weather, as otherwise the plant may be neglected. If possible, suitable gear should be provided at ground level.

Accessibility.—If it can be conveniently arranged that all parts which it is desirable should be kept under observation (such as the apportioning gear) are at, or near, ground level, or in some readily accessible place, the plant will necessarily be kept under closer observation than if placed high up, or in some out-of-the-way corner. In any case safe ladders should be provided for reaching all parts requiring attention. Flat step ladders are preferable, and hand-rails should be provided.

Short Circuiting.—Reaction and settling tanks should be of such shape and dimensions that the water will not be liable to take a short cut across from the inlet to the outlet, and thus render a portion of the tank useless.

If tanks are of large area this is almost sure to occur. Where a tank of large area is the most convenient, it should be divided off into two or more compartments, the water passing through them either in series or in parallel. Vertical cylindrical tanks are the best for avoiding short circuiting, and are to be recommended where in other respects they are equally convenient.

Direction of Flow.—To obtain the best results water should not pass through the settling tanks in a downward direction, otherwise the precipitate, which has a natural tendency to fall, will not be separated from the water as quickly as possible. If the water passes downwards, not only does it follow the downward tendency of the sediment and thus reduce the rate of separation, but on leaving at the bottom of the tanks the water may tend to carry along with it some of the mud already deposited there.

Sludging.—All settling tanks should be provided with convenient and effective sludging arrangements, otherwise mud will collect to such an extent as to interfere with the speedy clarification of the water. In this connection it may be said that, though it does no harm to keep *settling* tanks free from deposit, an exception must be made in those cases where one tank acts both as *reaction* and *settling* tank as in the case of the vertical cylindrical plants. When reagents and water have mixed, the chemical reactions are hastened if the water is allowed to agitate and mix with old precipitate, and larger particles of precipitate are formed which separate more quickly from the water. In such plants, therefore, it is not advisable to keep the bottom of the tank too clean, though, of course, undue accumulation of mud should be prevented by the regular use of the sludging gear.

Filter Area.—The filter or filters should be of such area that the water passes through at slow speed. As a rule wood-fibre filters are used, and if worked at too high a speed they are not effective. Even when two or more are used in series it is very desirable that *each* should be of ample area. Too much reliance must not be placed on wood-fibre filters. Though, if properly used, they are usually satisfactory in softening plants, unless the fibre is carefully packed (especially in the corners), and cleaned sufficiently often, they are liable to allow precipitate to pass through. If they are required to do much work, frequent cleaning will be necessary, and this is usually found very inconvenient. Provided the settling capacity is great enough, little work is left for the filters, and they may work for months at a time.

Water should always pass through wood-fibre filters in an upward direction.

Sand Filters.—In cases where the most efficient filtration is needed, a sand filter following the wood-fibre is an advantage. This may sometimes be necessary where perfect elimination of oil is required, and where the plant does not provide sufficient time for the sedimentation, or provide means for trapping any greasy scum. The oil being entangled in the precipitate, it is important that this should be efficiently removed.

Sand filters, when dirty, instead of passing imperfectly filtered water, choke up, and so demand the attention which they might not otherwise receive.

Storage.—In almost all cases a softened water storage tank is desirable, in order that a supply of treated water may be available in case of a hitch in the working of the plant necessitating temporary stoppage. The size of the storage tank must depend on the working conditions in each particular case, but frequently a storage sufficient to supply small requirements during the night is put down, so that the plant need only be worked during the day.

Painting and Housing.—All external parts of the plant should be well painted to protect them from rusting, and it is always an advantage if the whole plant can be erected indoors. Failing this, housing the working parts is desirable, and usually results in the plant being kept cleaner, and receiving closer attention.

CHAPTER VI

PRIMING

UNDER certain conditions, water is carried over with the steam from boilers. This is due to the steam bubbles, instead of bursting on the surface of the water, rising unbroken into the steam space and more or less filling it, some being carried right into the steam pipes. This is generally known as *priming* or *foaming.*

Priming may well be subdivided into two classes; "Foaming" and "Bumping."

Foaming may be considered to be the partial or complete filling of the steam space with a froth somewhat resembling soap lather, composed of numerous comparatively small bubbles of steam. In other words, it is too free ebullition. This may occur when :—

(*a*) The nature of the water is such as to form a froth when violently agitated. Soap or certain other organic matter such as Irish Moss may do this.

(*b*) There is too free a liberation of steam bubbles due to small particles of matter, suspended in the water, acting as nuclei for the formation of steam bubbles. Up to a certain point these suspended particles may be useful in causing smooth ebullition, but when present in excessive quantities, small bubbles of steam may be liberated so freely, and in such large numbers, as to produce a foam somewhat similar to that due to soap.

Bumping.—This term may be used to imply intermittent or explosive boiling, or, in other words, retarded ebullition. If the condition of the water is such as to interfere with the free formation or liberation of steam bubbles, the water may be quiescent for a short time and become superheated, and then large steam bubbles are likely to be formed suddenly and rise to the surface, where they may burst with sufficient violence to throw some of the water into the steam space, and possibly into the pipes. This may occur :—

(*a*) With certain solutions, such as caustic soda, which have a great tendency to boil intermittently in this manner.

(*b*) If a film forms on the surface of the water, thereby preventing the free liberation of steam. This is particularly noticeable in laboratory work when a highly saline solution is being evaporated. A point is reached when a very thin crystalline film is formed on the surface of the liquid, preventing or retarding evaporation. The liquid then becomes superheated until the vapour pressure is sufficient to burst through the skin, when the steam will be liberated with violence. A film of oil on the surface may cause a similar result.

Probably one of the most frequent causes of priming is the presence in the water of light particles of solid matter in suspension. It is a regular practice in laboratories, when a liquid is liable to boil explosively, to introduce small pieces of pumice, platinum wire, capillary glass tube, or a very small quantity of zinc dust. Such things act as nuclei for the formation of the steam bubbles, which then form freely and continuously, and the liquid boils quietly. Whatever is used for the purpose, it is important that it be introduced carefully, otherwise, at the moment of introduction, the liquid is very likely to froth up, and possibly overflow the containing vessel. In the case of zinc dust this effect is most pronounced, and if much be added the foaming will continue to an undesirable extent.

We will now consider how the foregoing remarks apply in the case of boilers. Finely divided suspended matter is likely to act similarly to zinc dust, the finer the particles, and the greater their number, the greater will be the foaming effect. Some particles will have only a useful effect in assisting free ebullition, while others may cause it to be excessive and lead to priming. Magnesium salts, when precipitated in a boiler, are apt to form a light floury deposit, which is probably the most likely substance to act in this way. Up to a certain extent this may be beneficial, but in excess it will very probably lead to undue foaming. The light floury magnesia precipitate is usually formed when magnesium salts predominate, and may therefore be present where imperfect softening has resulted in the removal of most of the calcium salts while leaving the magnesia.

Again, as already stated, at the moment of introduction of substances which act as nuclei, sudden foaming usually results. Such a condition may readily occur in boilers under certain circumstances. If a water be irregularly treated, either in a softening plant, or by the admixture of chemicals without a softener, the water in a boiler may contain either permanent hardness, or an excess of the chemical used. If the water in a boiler contains permanent hardness and an excess of soda enters, or *vice versâ*, a sudden precipitation occurs which may act similarly to the introduction of zinc dust, and have much the same result—namely, excessive foaming. This probably has a bearing on the much discussed question as to whether excess of sodium carbonate causes priming.

There is no doubt that boilers containing an excess of sodium carbonate do frequently prime, yet the author has had considerable experience with a number of boilers regularly working with several hundred grains of sodium carbonate and caustic soda per gallon, which gave no trouble through priming. Moreover, those which prime do not usually do so continuously, as might be expected were the soda to blame. Probably, in such cases, the trouble is due to permanent hardness entering the boilers and resulting in the sudden formation of a precipitate. This may readily be demonstrated by boiling a solution of sodium carbonate, and then adding a solution of any salt (such as magnesium sulphate) which will form a precipitate. Immediate foaming will result.

A similar effect may be obtained by introducing a small piece of hard scale, say a calcium sulphate scale, into a boiling soda solution. In this case the disintegration of the scale produces the fine particles necessary to cause foaming. This may often occur in boilers where there is old scale on the plates, and where an excess of soda is used in the treatment of the feed water, pieces of scale becoming detached and disintegrated.

It is a common complaint, when softening plants are installed, that the softened water *causes* priming. It may certainly *lead* to priming by loosening old scale, but had the boilers been thoroughly cleaned before commencing treatment there would probably have been no trouble. It may be argued that this explanation is incorrect, as many boilers are thus fed without ill-effect although they contain scale, but it must be remembered that moderate foaming is rather beneficial than otherwise, and may therefore occur in many cases unnoticed, while in others, owing to the different character of the scale, the greater excess of soda, or other circumstances, excessive foaming may result.

Mineral precipitates in a boiler are not the only finely divided substances liable to cause foaming. Organic matters in suspension are quite likely to lead to such results if present in sufficient quantity, especially those of a more or less colloidal nature.

Soap occasionally obtains access to feed water supplies and causes foaming, while in other cases it may actually be formed in boilers by the saponification of fixed oils by soda.

Bumping probably occurs less frequently in boiler practice. The concentration of caustic soda necessary to lead to bumping is not usually reached, and only very excessive concentration of salts would lead to the formation of crystalline skin on the surface of the water, but it is quite possible for either condition to be present. An oily scum is more likely to be present, but even this is not common now-a-days, most steam users avoiding the introduction of oil in such quantities as would bring about such a result.

It is claimed, on very good authority, that, in locomotive boilers, the soluble salts, sodium carbonate, sulphate, and chloride, may cause priming when present in such low concentration as 250 grains per gallon, or even, in very sensitive types of boiler, 160 grains. Though high concentrations of such salts will undoubtedly increase the surface tension of the water sufficiently to cause priming, it seems almost incredible that such weak solutions as those mentioned can do so, and the author cannot help thinking there may be some contributory cause. It is certainly a fact that hundreds of stationary boilers, of all types and sizes, working under widely differing conditions, give no trouble even with several thousand grains per gallon of saline matter.

A slight tendency to foaming may be intensified by the working conditions and become objectionable priming. Such aggravation may result from sudden excessive demands on the steam, as in the case of winding engines at collieries, and other engines which work intermittently; also where steam is used for boiling water as in the case of laundries. If a number of washing machines require steam at the same time, the supply may not be equal to the demand. In all such cases, excessive demand results in a sudden reduction of the steam pressure in the boilers, and the boiling point of the water being thus lowered, the water becomes superheated, and will boil vigorously until the balance is restored. If the tendency is already there, priming is likely to result. It is most important, where sudden demands on the steam are unavoidable, that boilers with large steam spaces be used, or, as an alternative, steam reservoirs installed.

Having discussed the most likely causes of priming, it will be well to mention the best methods of avoiding it.

(1) Ample steam space should be provided, especially where sudden demands are unavoidable.

(2) Soapy water should be carefully excluded, and also oil, especially if it contains fixed oil which may saponify and thus form soap.

(3) Excessive accumulation of any saline matter, especially caustic soda, should be avoided by the regular use of the blow-off cock, and sufficiently frequent emptying of the boiler.

(4) Prevention, if necessary, of undue accumulation of fine particles in boilers—

(*a*) by avoiding suspended matter in the feed water, filtering if necessary ;

(*b*) by adopting efficient softening if the water is hard, taking special care to reduce magnesia to the lowest possible limit ;

(*c*) by, in any case, removing as far as possible (by the regular use of the blow-off cock) such deposit as cannot be avoided.

(*d*) Whatever treatment, if any, be adopted, should be as uniform as possible, in order that permanent hardness and soda entering the boiler alternately may be prevented.

In connection with priming it must be borne in mind that it may be not only a source of danger by water-logging the steam pipes, and thus allowing water to enter the engine cylinders, but it is conceivable that priming in the tubes of a water-tube boiler, especially where the tubes are of small diameter, might lead to the temporary partial emptying of one or more tubes, and consequent over-heating.

It may be well, at this point, to make a few remarks as to the blowing off of boilers for the purpose of preventing undue concentration of soluble salts, or the accumulation of mud.

If it is only necessary to keep down the concentration

of soluble constituents, opening of the blow-off cock once, or perhaps twice a day, may be sufficient, reducing the water level each time by, say, from 1 to 6 inches, according to the requirements of the particular case.

On the other hand, if it is necessary to prevent the accumulation of mud, the blow-off cock should be opened, for a few seconds only, several times a day. The effect, when the blow-off cock is opened, as as follows :—

While working, a more or less uniform layer of deposit will probably be formed on the bottom of boilers. When the blow-off cock is opened, the deposit over it, and in the immediate vicinity, first runs out, but afterwards water containing only a comparatively small proportion of sediment leaves, and the remainder of the deposit on the bottom stays behind. After closing the cock the deposit gradually levels itself again over the blow-off pipe, so that after a time, a further quantity may be removed as before. Opening the blow-off cock several times for a few seconds has therefore a much greater effect in removing deposit than blowing off the same amount of water in one operation. It must not be overlooked, however, that frequent opening and closing of the cock leads to wear and tear, and it is for the engineer in charge to decide how often it is necessary in any particular case.

In certain cases it may be desirable to arrange for continuous blowing-off of a small proportion of the water, suitable means being adopted to ensure that the water thus removed is collected from the whole of the main surface on which the mud is deposited. The author strongly deprecates the method, recently advocated by Kestner, of returning the blown-off water to the boiler, after allowing the mud to deposit. The slight loss of heat due to reasonable blowing down is nothing as compared with the advantage gained by keeping down the concentration.

It is convenient, at times, to be able to calculate how many days a boiler can be worked, with a known amount of blowing down, before a certain concentration is reached, or what will be the maximum concentration possible with a certain amount of blowing down. The author's formulæ for calculating these are given in the Appendix.

CHAPTER VII

SCALE, GREASE, AND OVERHEATING

SCALE in boilers is so common that all who have anything to do with steam boilers are well acquainted with it, and very little need be said on the subject.

In most cases it is very difficult to offer a definite opinion as to the character of the scale likely to be formed from any particular water, as so much depends on working conditions. Where the hardness of a water consists almost entirely of calcium sulphate it is fairly safe to predict that the resulting scale will be of a hard nature whatever the conditions, but beyond this it is seldom safe to venture. Magnesium salts have a tendency to form a light floury deposit, but in combination with other constituents may produce a very hard scale. In any case the character of a scale may change after it is first deposited, owing to the temperature of the plates, and mud or soft deposit in boilers is frequently baked, and thus hardened, when boilers are emptied while hot.

There is no doubt that scale or deposit of any character will have some effect in retarding the passage of heat through the plates of a boiler, and it is often contended that it results in serious loss of heat. This is open to doubt, however, as, except in exceptional cases, the imparting of heat to the water may be only transferred from one part of the boiler to another, and the temperature of the gases entering the chimney will not be materially different in a moderately scaly boiler, from that in a fairly clean one. Where economisers are used it may be said that probably little loss of heat results from moderate scale in the boilers.

It must not be thought that if there is no loss of heat scale is unobjectionable. In the first place, the expense of scaling boilers is often a very serious item in the working costs. Again, it may be decidedly inconvenient to shut down boilers for the time necessary for proper scaling, and it is not uncommon for boilers to be allowed to get into a serious condition owing to the impossibility of temporarily dispensing with their services.

Another drawback to scale is the fact, referred to previously, that it sometimes assists corrosive constituents to be more virulent. Many very serious cases of corrosion occurring beneath scale may be found in boiler practice, and such wasting sometimes goes on unnoticed until a plate has become dangerously thinned, if not actually perforated.

A further and very important ill-effect of scale is the increase in the temperature of the plates (overheating) which will result to a greater or less extent, dependent on the character and amount of the scale, of the parts of the boiler on which it is deposited, and the rate of firing of the boiler. As a rule, in the absence of grease in the scale, the degree of overheating will not be such as to lead to definite evidence at the part or parts where it occurs, but may result in generally increased strains, and thereby possible leakages at seams.

Overheating becomes very much more pronounced when oil is present in the scale or deposit. In such cases local effects are very likely to be found, such as bulging or ovality of furnace tubes, bulging of shell bottoms, and bulging or hogging of the tubes in water-tube boilers.

It is surprising how little grease in scale may lead to distortion where the rate of firing is high, or where, for some reason the flames impinge on one particular part. As little as one-tenth per cent. of grease in scale or deposit may cause such results, whereas boilers fired more easily may work for years without trouble with a much greater proportion of grease.

An idea exists among many steam users that if the water is soft a small quantity of grease is of little moment. This is incorrect, as, with a certain proportion of oil in a water, the lower the amount of scale-forming matter the higher will be the percentage of grease in the deposit formed, and therefore the greater the risk of overheating troubles.

Grease cannot be totally removed from steam by mechanical means such as adopted in the various types of oil separators. These usually consist of devices for causing the steam to impinge on baffles and thus changing its direction, or for suddenly reducing its pressure. Undoubtedly such methods will effect the removal of a good deal of oil, but the more volatile portions of the oil will be in the form of vapour and will remain as such in the steam until it is condensed.

On the other hand, grease cannot be efficiently removed from water by simple filtration through cloths, coke, wood-fibre, or such like.

Where overheating troubles due to oil are experienced, it becomes necessary to eliminate it totally from the water, or at least to reduce it to a negligible amount. This may be accomplished by what is known as coagulation. A precipitate is produced in the greasy water, and the tiny globules of oil are thus enclosed in, or entangled by, the precipitate. The precipitate is afterwards removed by sedimentation and filtration.

The coagulation process is usually carried out either by softening treatment, or in a plant similar to a softening plant. If the water contains sufficient hardness (the amount necessary for the purpose depending on the proportion of oil), ordinary softening should efficiently entangle the oil, and, provided the precipitate be completely removed, the water should be suitable for use in boilers. The precipitate formed by magnesium salts is more effective for grease removal than that from calcium salts.

Where the water has not enough hardness to give a sufficiently large precipitate, it becomes necessary to introduce another coagulent, alumino ferric (commercial sulphate of alumina) being usually chosen. To some extent alumino ferric may be precipitated by the temporary hardness (if any) in the water, but usually a precipitant has also to be used, caustic soda being the best for the purpose. Care must be taken that the alumino ferric is completely removed, otherwise corrosion may result. The aluminium hydrate precipitate formed in these cases is of a gelatinous nature and is a very effective coagulent, much better than either lime or magnesia precipitates.

In plants using alumino ferric, it is important that it be introduced and distributed throughout the water before the precipitant (caustic soda) is added, otherwise the precipitate may all be formed in a comparatively small area, and the globules of oil in the remaining water may escape entanglement.

An electrical process is on the market for de-greasing water, and is useful where the necessary electric current is to be obtained sufficiently cheaply, as in the case of electric generating stations. It is known as the Davis Perrit process, and consists of passing the water between two steel or iron plates fixed parallel with one another and fairly close together. An electric current is passed between the plates through the water, iron being thereby dissolved from the anode and precipitated in the water, entangling the oil. The precipitate is afterwards removed by filtration.

By efficient de-greasing treatment it is usually possible to remove oil to such an extent that not more than 0·02 grain per gallon remains.

It is usual in cases where water from condensing plants is used for boiler feeding purposes that grease eliminating plants become necessary. Whenever possible, in such cases, whatever type of condenser be used, the steam should be passed through an efficient oil separator before being condensed, as the greater part of the oil is thereby removed, thus increasing the efficiency of the de-greasing process.

In the case of surface condenser waters required for boiler feeding purposes, it is almost always necessary to have some make-up water, and frequently the make-up is of such a character as to render softening desirable. Where this is so, the proportion of condensate to make-up usually varies more or less considerably, and the mixture is difficult to treat uniformly. The best plan is to have separate plants—one for softening the make-up, the other for de-greasing the condensed water. In most cases this is considered to be too expensive, or to require too much supervision.

An alternative plan is to have only one plant, but the reagent gear must be operated by the make-up water only, the condensate and make-up mixing when the latter has left the tipping-bucket or other measuring device. The reagents introduced will need to be those required for the treatment of the hard water, *plus* alumino ferric and a suitable proportion of precipitant. It is essential for the success of this method that the supply of make-up water shall be continuous, otherwise some condensate will pass through the plant untreated. The condensate should also be supplied as uniformly as possible.

As the hardness of the water cannot be reduced below a certain degree, no matter how low it is originally, it follows that the plan of having separate plants will result in a much less sedimentary feed supply. Supposing the hardness can be reduced to two degrees, and that ten per cent. of the total water consists of make-up, then with separate plants there will be ten per cent. of water with two degrees of hardness, and 90 per cent. with none, or one-fifth of a degree in the mixed supply. If one plant is used the whole supply will only be reduced to two degrees.

CHAPTER VIII

BOILER CONTROL

In the preceding chapters various remarks have been made respecting the concentration of soluble salts, accumulation of mud, and the rate of firing of boilers. A few remarks on the working of boilers, so far as the water is concerned, may perhaps be helpful.

Though the loss of heat through blowing off water is not as serious as is frequently stated, no careful engineer is willing to waste even a small quantity of heat unnecessarily. It becomes a question of how much blowing down is desirable or economical in each particular instance. Obviously no hard and fast rules can be given, as every case must be considered in the light of the composition of the feed water, the type of boiler, the working pressure, the rate of firing, and the demands on the steam. The questions can only be decided by the engineer in charge, who is in the position to note the effects and make his own calculations and deductions.

The degree of concentration of water in a boiler is generally considered by engineers, either in terms of ounces of saline matter per gallon, or in the marine system of a certain number of "*thirty-seconds*." A "*thirty-second*" is the concentration equivalent to the normal salinity of sea water, which contains approximately one thirty-second of its weight of salt—*i.e.*, five ounces, or about 2,200 grains, per gallon.

The density of the water in boilers is generally ascertained by drawing a sample from the gauge cock, after first flushing out and closing the steam cock, and testing with a salinometer, but, where frequent observation of the density is important, a device known as the " Kek " densimeter is most useful, the density at any time being indicated by the difference in level of the water in the two gauge glasses. The principle adopted is to balance a column of distilled water against a column of the water in the boilers, both being at the same temperature. A condensing device is fitted to the top of one of the gauge glasses, and by this means a constant stream of distilled water trickles into the gauge glass. The lower end of the same glass is connected to an internal pipe, which reaches nearly to the bottom of the boiler. The whole of this gauge glass and internal pipe, being constantly filled with distilled water, the level in the glass is higher than that in the other glass, the difference in level being an indication of the density of the boiler water. A scale is provided, as shown in Fig. 18, from which the actual density can be ascertained.

Though some boilers can be worked satisfactorily up to a concentration of two-and-a-half thirty-seconds (5,500 grains per gallon), it is generally desirable to avoid anything approaching this. The higher the salinity of the water, the greater will be the risk of corrosion, priming, overheating, and strains, and it is good practice to keep down the concentration as low as possible consistent with economy.

The factors to be considered when deciding how much and how frequently boilers should be blown down, and how often they should be emptied completely, will include :—

(1) The character of the works. In some works it is

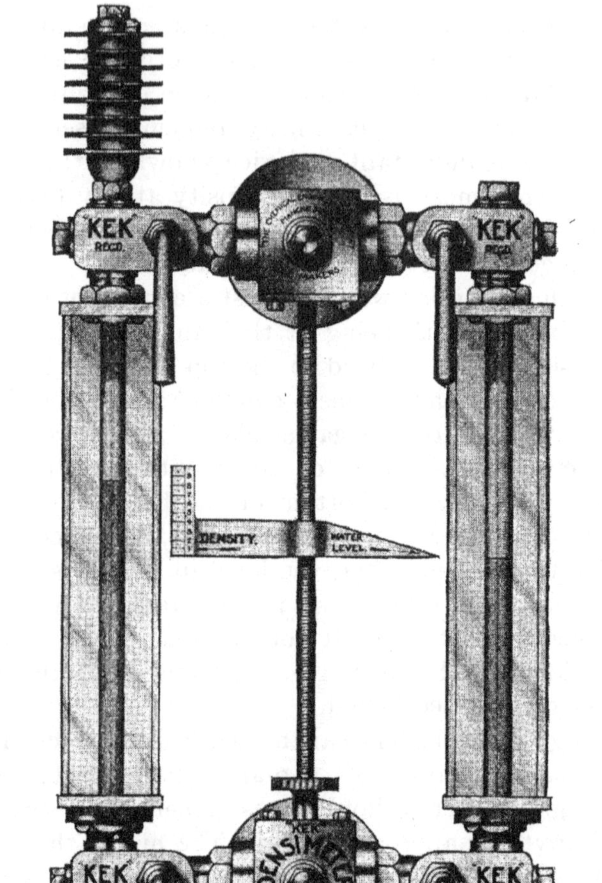

Fig. 18.—" Kek " Duplex Densimeter.

possible to empty boilers at any week-end, but in others it may be almost impossible to do so except at holiday times.

(2) The facilities for blowing off. Some boilers can be blown down at any time without inconvenience, whereas others cannot be blown down conveniently during working hours.

(3) The design and position of the blow-off cock. Some are readily accessible and easily opened, others being inconveniently placed and difficult to open.

(4) The characteristics of the water. Feed water containing a high proportion of saline matter or scale forming constituents, may necessitate the blowing-off of large quantities of water or blowing down frequently, and the boilers will need to be emptied fairly often. Other waters are so pure that boilers may be worked for considerable periods and with very little blowing down.

Formulæ are given in the Appendix which will be found useful when considering this important question. If it has been decided that a certain concentration should not be exceeded, the number of days the boiler may be worked before reaching this can be ascertained, or a calculation can be made of the quantity of water which must be blown off daily (or periodically), in order that the desired concentration shall not be reached before a certain length of time. It is also possible to calculate what will be the approximate maximum concentration attainable with a given degree of blowing down. The table of capacities of horizontal cylinders filled to different levels will be helpful when making these calculations.

As a general guide it is suggested that an endeavour

should be made to keep the saline constituents below 1,000 grains per gallon, where this can be accomplished economically. Nitrates should be specially considered in the case of high-pressure boilers, and the maximum concentration should be such that the nitrates should not exceed, say, 100 grains per gallon, and even less if corrosion attributed to them still occurs.

When mud is found to accumulate unduly in the boilers, instead of blowing down boilers to a certain extent once daily, the same amount of water should be removed by opening the blow-off cock several times a day, for a few seconds each time. The effect of this has been explained on p. 80. In special cases some system of continuous blowing down may be adopted with advantage.

The rate of firing of boilers is of great importance, and should be kept as low as possible. The higher the rate of firing the higher will be the temperature of the plates, and, consequently, not only will the liability to corrosion increase, but, particularly where traces of grease are present in the scale, the greater will be the tendency to overheating and bulging of the furnaces or tubes. Many boilers, which have worked satisfactorily with low rates of firing, have had serious overheating troubles when the rate of firing has been raised to twenty-five or more pounds of coal per square foot of grate area. Where increased output calls for more steam, it is policy to consider whether additional boiler plant cannot be installed in preference to unduly increasing the rate of firing.

As far as possible boilers should be allowed to cool down naturally before emptying. Sudden cooling tends to cause straining, and emptying while hot will result in baking the scale or mud, making them more difficult to remove when cleaning.

When cleaning boilers, all pittings and crowded areas should be scraped free from oxide, otherwise there will be great liability to continued wasting.

Any sign of red colouration in the water, as seen in the gauge glasses, should lead to immediate investigation of the cause, as an unexpected change may have taken place in the composition of the feed water, or pollution of some sort may have occurred.

CHAPTER IX

METHODS OF ANALYSIS

For the convenience of chemists engaged in other branches of the profession who may be called upon to make an occasional analysis, a description of the methods used by the author for a number of years, and proved to be thoroughly satisfactory, will probably be useful. In order to make them as helpful as possible they are given in some detail, and the size of apparatus found most convenient is given. Hints are also included as to the best order of commencing the various determinations. The quantities of water used are in some cases less than those recommended by many chemists, but are sufficient for all practical purposes, except in special cases, when larger quantities may be used at the discretion of the analyst.

Many attempts have been made to devise quicker methods, but the results obtained by them are not usually sufficiently accurate or reliable, and as a rule, little, if any, time is saved. A complete analysis of a sample of water can be comfortably carried through in an eight-hour day, with some spare time between operations. Where a number of samples are to be analysed they are most conveniently done in batches of four, which can be completed in two days.

The strengths of the reagents will be given in another chapter.

In order to avoid delay, the following order for starting the various estimations may be found useful when commencing analyses. The determinations of dissolved gases, Carbonic Acid and Oxygen, should first be commenced if these are required. If an estimation of Suspended Matter is desirable, the sample should then be well shaken and a portion measured off for the purpose, the remainder, or as much as necessary, being filtered for the other determinations. Should it be unnecessary to estimate the Suspended Matter, it may be possible to decant the clear water without disturbing the sediment. Next measure out the quantities for Lime and Sulphates and commence the evaporations after acidifying. When these have been started, commence the Total Solids and Chlorine quantities of the first sample, and determine its Alkalinity. The pipette can then be rinsed with the second sample, and the same determinations put in hand, the other samples being thus started in turn. Nitrates should not be commenced till the Chlorine determinations have shown how much water is to be used. Ammonia may be estimated at any convenient opportunity.

Free Carbonic Acid should be estimated as soon as the bottle is first opened. If suspended matter is present it should be allowed to subside before opening the bottle, so that the quantity for this estimation may be decanted rather than filtered.

Decant 100 c.c. into a stoppered 100 c.c. cylinder, and compare against 100 c.c. of well boiled and cooled distilled water in a similar cylinder. Add ten drops of phenol phthalein solution to the blank, and then Standard Sodium Carbonate solution from a 10 c.c. burette, one drop at a time, stoppering and gently mixing between each addition. When a faint permanent pink colour is obtained, note the burette reading. Test the sample similarly, but with this the standard solution can usually be introduced several drops at a time to begin with. Continue till a permanent tint is obtained about equal to that in the blank.

The number of c.c. of Standard Sodium Carbonate required by the sample, less that needed by the blank = grains of Free CO_2 per gallon.

Suspended Matter need not as a rule be estimated, as it will usually vary so widely in samples from the same source, but if required 1,000 c.c. (less if much present) of the thoroughly shaken water is filtered through a $12\frac{1}{2}$ cm. folded filter which has been washed with water, dried, and weighed. Wash, dry, and weigh again.

Weight of Suspended Matter in grammes from 1,000 c.c. of water $\times$ 70 = grains per gallon.

Lime, Magnesia, and Silica.—Evaporate 250 c.c. to dryness with about 5 c.c. of hydrochloric acid in a 6-inch porcelain dish, finishing the evaporation on a steam hole. Cover with a clock glass and gently heat over a small flame till any organic matter is thoroughly charred. When cool, add 5 c.c. hydrochloric acid, and 5 c.c. water, cover, and simmer gently for about two minutes. Filter through a $5\frac{1}{2}$ cm. ashless paper ($1\frac{1}{2}$-in. funnel) into a tall beaker (3 in. $\times$ $1\frac{1}{2}$ in.), rubbing the last traces of silicious matter from the dish with a clean finger.* Wash the filter paper and silica three times with hot water. The filtrate, which should usually be about

* Occasionally when much calcium sulphate is present it may not be dissolved completely in the 10 c.c. of the diluted acid. In such cases it is readily noticeable, being a heavy, white powder, easily distinguished from the almost invisible gelatinous silica. When this is found, the solution should be decanted through the filter, and a little more water and a drop or two more acid used to complete the solution of the calcium sulphate.

30 to 35 c.c., and contains the lime and magnesia, together with iron, alumina, and manganese if present, may then be removed. Complete the washing of the filter, and ignite and weigh the silica.

Weight of Silica in grammes × 280 = grains per gallon.

To the filtrate from the silica add two or three drops of bromine water (more if much manganese is expected) and then ammonia (·95) till the colour of the bromine is destroyed and there is a slight smell of ammonia. About 5 c.c. of ammonia will be required, and only about two drops excess should be used. Cover the beaker with a watch glass, and heat *in* a steam hole for two or three minutes. If any precipitate forms, the beaker must be allowed to remain in the steam hole till the precipitate has satisfactorily coagulated, after which it is filtered through a 7 cm. ashless filter ($1\frac{1}{2}$ in. funnel) and the precipitate dealt with for Iron, etc., as described later. Transfer the beaker of hot solution (or filtrate if there has been a precipitate) to the top of the bath, add 3 c.c. of ammonium oxalate solution drop by drop, while shaking or stirring with a glass rod (if stirred, the rod should not be allowed to touch the sides or bottom of the beaker). Stir or shake occasionally till the precipitate settles fairly quickly, and if much lime is present add a few more drops of ammonium oxalate to test if precipitation is complete. Keep the beaker on the bath for not less than half an hour, and not more than one hour, by which time the precipitate should settle quickly after shaking, and leave the supernatent liquid clear. Filter through a 7 cm. ashless paper, collecting the filtrate in a tall beaker (4 in. × $1\frac{3}{4}$ in.). Wash six times with small quantities of hot water, retaining only the first three washings. Ignite the precipitate and weigh as CaO. The final ignition may be for three minutes over a blowpipe, or five minutes over a No. 4 Meker burner if a platinum crucible is used. Longer ignition is necessary for specially large precipitates. In order to make sure that the lime is properly calcined, it should be tested, after being weighed, by slaking it with a few drops of water, and then adding hydrochloric acid.

Weight in grammes × 280 = grains of CaO per gallon.

Cool the filtrate from the calcium oxalate, and transfer it to a conical flask (150 c.c. Registered Pattern). Add slowly, while shaking the flask, 3 c.c. of a saturated solution of sodium hydrogen phosphate, and then strong ammonia (·880) equal to about one-third the bulk of the solution, shaking it during the addition. Cork the flask with a rubber stopper. If possible allow to stand overnight after shaking it till precipitation has commenced. If required sooner it may be shaken for fifteen minutes, and allowed to stand not less than one hour. Filter through a 7 cm. ashless paper (preferably not too soft in texture), and wash six or eight times with small quantities of washing ammonia (see chapter xii.). Dry in a steam oven, detach the precipitate as completely as possible from the paper, ash the paper in a platinum crucible, and then introduce the precipitate. Cover the crucible with a lid, and heat over a bunsen, first with a flame barely touching the crucible, then gradually increasing the flame until full on. Remove the lid, and continue the ignition till all carbon is burnt away. Cool and weigh as $Mg_2P_2O_7$.

Weight of $Mg_2P_2O_7$ in grammes × 101·43 = grains of MgO per gallon.

Iron, Aluminium and Manganese.—If the ammonia precipitate obtained from the lime quantity is very small it will in most cases be sufficient to ignite and weigh it, considering it as iron, aluminium, or manganese, or a mixture, according to its colour when precipitated, which is usually a fairly accurate guide. Should it be considered necessary to estimate each constituent, dissolve the precipitate through the paper with hot dilute hydrochloric acid, and pour it slowly into a boiling solution of pure caustic potash in a nickel dish, making sure that the caustic potash is decidedly more than will be necessary to neutralise the acid. Boil, filter, and wash the precipitate, which contains the iron and manganese. Acidify the filtrate with hydrochloric acid, and precipitate the alumina in the boiling solution, in a 5 in. deep form porcelain dish, with the least possible excess of ammonia, boil, filter, wash, and ignite the precipitate over a blowpipe or Meker burner, and weigh as Al_2O_3.

Weight in grammes × 280 = grains of Al_2O_3 per gallon.

Dissolve the precipitate of iron and manganese in hydrochloric acid, add bromine water, and reprecipitate with ammonia while boiling in a porcelain dish. Boil, filter, wash, ignite over a blowpipe or Meker burner, and weigh as mixed oxides of iron and manganese.

Weight of precipitate in grammes × 280 = grains of Fe_2O_3 and Mn_3O_4 per gallon.

Estimate Iron in a separate quantity of water by the colorimetric method as follows. Boil a suitable quantity of water (not exceeding 50 c.c.) for about one minute with three or four drops of hydrochloric acid and two drops of strong nitric acid. (If 50 c.c. of water has been used, evaporate to about 40 c.c.) Cool, add 6 c.c. of potassium sulphocyanide solution (10 per cent.) and 2 c.c. dilute hydrochloric acid, and dilute to 50 c.c. in a Nessler cylinder. Into a similar cylinder put 6 c.c. of potassium sulphocyanide and 2 c.c. dilute hydrochloric acid, and dilute to rather less than 50 c.c. Add, from a burette, Standard Ferric Solution till the colour of the sample quantity is exactly matched, completing the dilution to 50 c.c. when nearly sufficient Ferric solution has been added.

c.c. of Ferric solution $\times$ 7 $\div$ c.c. of water taken = grains of Fe per gallon.

Fe $\div$ 0·7 = grains of Fe_2O_3.

The quantity of water taken should be such as will require not more than about 2 c.c. of Ferric solution otherwise the colour will be too deep for accurate comparison. A suitable quantity can be decided upon by making a qualitative test on about 10 c.c. in a test tube.

The difference between the mixed Fe_2O_3 and Mn_3O_4, and the Fe_2O_3 separately determined = Mn_3O_4.

Sulphuric Anhydride.—Evaporate 500 c.c. in a 7 in. porcelain dish, acidifying with about eight drops of hydrochloric acid. When reduced to about 40 to 50 c.c. rinse into a 6-oz. conical beaker with hot water. The total volume should not be more than 70 c.c. unless much sulphate is present. Raise to boiling, and add about 5 c.c. barium chloride solution, shaking all the time. Boil for a few seconds, and then add a little more barium chloride to test if precipitation is complete. When the sulphates are all down, boil for a few minutes, and allow to stand on a steam hole for at least two hours. Filter through a 7 cm. paper of fine texture, wash well with nearly boiling water, ignite, and weigh as $BaSO_4$. If the precipitate is large it should first be washed several times by decantation, boiling up each time.

Weight of $BaSO_4$ in grammes $\times$ 48·016 = grains of SO_3 per gallon.

Total Dissolved Solids.—Evaporate 70 c.c. to dryness in a platinum dish on a steam hole, heat in an air oven at 270° Fahr.* till constant in weight. It is usually advisable to heat for about three hours before the first weighing. Always closely cover the dish while cooling in the desiccator.

Weight of solids in milligrammes = grains per gallon.

It is sometimes advisable to ignite the solids *very gently* after weighing, and weigh the Inorganic Solids.

Chlorine.—Evaporate 70 c.c. in a $3\frac{1}{2}$ in. porcelain dish over a steam hole, till only two or three drops remain. Cool, add one drop of potassium chromate solution, and titrate with Standard Silver Nitrate till the slightest permanent brown tint is obtained, which remains after stirring and washing down, by means of the solution, any chlorides which remain on the sides of the dish. Deduct 0·05 c.c. from the observed reading, for excess of silver nitrate used in obtaining the brown tint.

If alkaline sodium salts are present, as indicated by the alkalinity being greater than the total hardness, they may interfere with the titration. In such cases, all but about 2 grains of the alkalinity should be neutralised by the addition of a suitable measured quantity of N/50 Sulphuric Acid before evaporating the water for this determination.

Waters liable to become acid when evaporated, such as those containing iron, should be stirred with a few milligrammes of pure calcium carbonate, evaporated nearly to dryness, filtered, and again evaporated down.

c.c. of Standard Silver Nitrate = grains of Cl per gallon.

If the Chlorine is over 20 grains per gallon, less than 70 c.c. of the sample should be used, whilst if it is particularly high it may be advisable to estimate it gravimetrically.

* If dried at a higher temperature Nitrates may be decomposed.

Alkalinity.—Choose, and reserve for alkalinity determinations, two deep 5 in. porcelain dishes which match each other in colour. Into one, used for purposes of comparison, measure 0·5 c.c. methyl orange solution with a pipette, and run in 0·4 c.c. N/50 Sulphuric Acid from a 50 c.c. burette, and afterwards refill the burette. Into the other basin measure 70 c.c. of the sample, and add a few drops phenol phthalein solution. If no red colour is produced the whole of the Alkalinity is present as Bicarbonate, and the titration may be conducted with methyl orange as directed below. If a red colour is produced, slowly add N/50 Sulphuric Acid from the burette, the jet being near the surface of the water, stirring with as little agitation as possible. The acid required to destroy the colour completely indicates the alkalinity to Phenol Phthalein.

Continue the titration after adding 0·5 c.c. methyl orange solution from a pipette. (Stirring may now be more vigorous.) As soon as the least change from yellow to pink occurs, note the burette reading, and add to the comparison basin 70 c.c. of distilled water plus a further quantity about equal to the volume of acid used in the titration. Continue the titration till the colour of the sample matches that of the blank. As the colour of the blank was produced by 0·4 c.c. of Standard Acid, it follows that 0·4 c.c. excess has been used in the titration, so that 0·4 c.c. must be deducted from the burette reading to obtain the amount required by the water.

N/50 Acid required in c.c. = total Alkalinity in grains per gallon, expressed in terms of Calcium Carbonate.

If there is no alkalinity to Phenol Phthalein all the alkalinity is present as *Bi*carbonate.

If the Phenol Phthalein Alkalinity is *not more than half* the Total, on multiplying it by 2 the Alkalinity as

Carbonate is obtained, and the difference between the Carbonate and the Total represents *Bi*carbonate Alkalinity.

If the Phenol Phthalein Alkalinity is *more than half* the Total, the difference between it and the Total Alkalinity, multiplied by 2, equals the Carbonate; and the difference between the Total and the Carbonate is Caustic Alkalinity.

Nitric Anhydride.—If the water does not contain more than about one grain per gallon of chlorine, 70 c.c. may be used for the determination. Where more chlorine is present a smaller quantity of the sample should be taken, such as will contain not much more than 1 mg. of chlorine. In cases where the chlorine is particularly high it is necessary, before commencing the determination of nitric anhydride, to add a solution of a weighed quantity of silver sulphate sufficient to precipitate nearly, but not quite, all the chlorine. After heating till the precipitated silver chloride has thoroughly coagulated, the solution is filtered, and the estimation proceeded with.

Evaporate a suitable quantity of the water * to dryness over a steam hole in a $3\frac{1}{2}$ in. porcelain dish *kept specially for the purpose.* (If the dish has been used for other work where nitrates are present, such as the determination of chlorine, a high result may be obtained owing to nitrates retained in the glaze.) Waters liable to become acid when evaporated should first be stirred with a few milligrammes of calcium carbonate, evaporated nearly to dryness, filtered, and then evaporated to dryness. Cool, and add from a pipette approximately 2 c.c. of phenol disulphonic acid, thoroughly wetting all the residue in the dish with it,

* Do not use the total solids quantity as some of the nitrate may have been decomposed on drying at 270° Fahr.

using a glass rod to assist in so doing. When all the residue is seen to have been thoroughly disintegrated, dilute with about 50 c.c. of cold water, add an excess of caustic potash (50 per cent. solution convenient), which in the presence of nitrates produces a yellow colour. Dilute with water to 100 c.c. in a measuring flask, and filter if necessary ($12\frac{1}{2}$ cm. folded filter). Compare the colour against standards (prepared as below) in 50 c.c. Nessler cylinders, diluting it quantitatively till it exactly matches one of the standards. In order to avoid wasting the standards, it is better to choose one paler than the sample. The sample quantity can be carefully diluted, till 50 c.c. matches the chosen standard, and the amount of dilution required noted.

If 50 c.c. of the sample has been used, and 50 c.c. of the yellow solution obtained after treatment and dilution to 100 c.c., requires diluting to 68 c.c. in order to match the 0·05 grain nitrogen standard, then the nitrate present in 50 c.c. of the water is equal in amount to that in 70 c.c. of a water with $\frac{68}{50} \times 0{\cdot}05 = 0{\cdot}068$ grain nitrogen. In other words, the sample contains $\frac{7}{5} \times 0{\cdot}068 = 0{\cdot}095$ grains of N, or $0{\cdot}095 \times 3{\cdot}855 = 0{\cdot}336$ grain N_2O_5 per gallon.

Owing to the presence of organic matter, the yellow colour is in some cases of a slightly different tint from the standard, but a sufficiently accurate comparison can usually be made. In such cases, if greater accuracy is necessary, the nitrates may be determined by the nitrometer method. (See next page.)

Some filter papers appreciably affect the colour of the alkaline solution, and the particular brand used for filtering it should be tested in the first place by comparing the colour of the solution after filtration, against a portion which has been decanted.

To prepare the nitrate standards, take 20 c.c. of the Standard Potassium Nitrate solution, which will contain 2 mg. of Nitrogen. Evaporate to dryness, treat with 2 c.c. phenol disulphonic acid, dilute, make alkaline with caustic potash, and dilute to 200 c.c. This will be a 1 grain Nitrogen Standard. Dilute portions of this to make other convenient standards, such as 0·5, 0·2, 0·1, 0·05, 0·04, 0·03, 0·02, 0·01 grain, keeping each in a 100 c.c. stoppered bottle. These standards appear to keep indefinitely, but as the caustic potash attacks the glass they will occasionally need filtering, and it is wiser to make up an entirely fresh set from time to time.

Nitrates by the Nitrometer Method.—For this purpose evaporate 250 c.c. (less if much nitrate is present) in a 6-inch porcelain dish till only about two or three cubic centimetres remain, filter through a 5½-cm. paper into a 3-inch porcelain dish, washing the filter two or three times with small quantities of water. Evaporate to about 1 c.c. and rinse into a Lunge's nitrometer (filled with mercury), making the volume of solution in the nitrometer just 3 c.c. Run in strong sulphuric acid (1·84) very carefully, to make the total volume 7·5 c.c. Expel the carbonic acid which will usually be evolved. Shake thoroughly and vigorously, so that the mercury mixes with the acid solution, and until no more gas is evolved when the solution is further shaken. Pour on to the mercury in the levelling tube a mixture of 3 c.c. of distilled water to which strong sulphuric acid has been added to make a total volume of 7·5 c.c. When the liberated gas is cold, adjust the level of the mercury in the two tubes, and read off the volume of Nitric Oxide (NO). Correct the volume of gas for temperature and pressure.

Corrected volume of gas in c.c. from 250 c.c. water $\times$ 0·175 = grains of N per gallon.

Nitrogen $\times$ 3·855 = Nitric Anhydride (N_2O_5).

In order to avoid the necessity of correcting the volume of NO for temperature and pressure, it is a good plan to keep a second nitrometer, with the stopcock specially well greased and tied in, charged with the NO liberated from 0·103 gramme of pure potassium nitrate, which will be equivalent to 4 grains of Nitrogen per gallon when working on 250 c.c. of water. Keep this standard beside the other nitrometer, and take a reading whenever a sample quantity is being read off.

c.c. of NO from 250 c.c. of the sample $\times$ 4 $\div$ c.c. of NO in the Standard tube = grains of N per gallon.

Ammonia should be tested for in every sample, by adding Nessler solution. If more than a faint yellow colour is produced the ammonia should be estimated.

To a suitable quantity of water (see below) add a few milligrammes of sodium carbonate, and a drop or two of 50 per cent. caustic soda, in order to precipitate the hardness. Filter if necessary, and measure out a suitable quantity into a 50 c.c. Nessler cylinder by means of a pipette, and dilute to the mark with water. Add 2 c.c. of Nessler solution and stir. In an exactly similar cylinder have nearly 50 c.c. of distilled water, add 2 c.c. of Nessler solution, and then add Standard Ammonium Chloride solution from a burette in suitable small quantities, until, after standing two or three minutes, the colour matches the sample quantity. When nearly sufficient Standard Ammonium Chloride has been added, the comparison quantity should be diluted with water, if necessary, till its volume is equal to that of the sample quantity.

c.c. of Ammonium Chloride solution required $\times$ 3·5 $\div$ c.c. of water taken = grains of NH_3 per gallon.

The quantity of water used should not exceed 50 c.c. for this determination, and should be such as will require not more than about 2 c.c. of Standard Ammonium Chloride to match the colour.

Oxygen.—It is sometimes desirable to determine the amount of oxygen in solution in water. If possible the test should be carried out, up to the point of titrating, as soon as the sample has been obtained; otherwise the sample should be kept as cool as possible, and in a well-filled stoppered bottle, and the determination commenced as early as possible.

To 250 c.c. of water in a stoppered flask or bottle add about 2 c.c. each of manganous sulphate solution, and potassium iodide-sodium hydrate solution, each solution being added from a pipette dipping slightly below the surface of the water. Stopper, and gently mix. Add about 2 c.c. dilute sulphuric acid, and mix thoroughly. After this the estimation may be continued when convenient. Titrate with N/40 Sodium Thiosulphate solution, adding a little starch solution towards the end of the titration. The Thiosulphate solution should be standardised by titrating with it 10 c.c. of accurate N/40 Potassium Bichromate, after adding 3 c.c. of potassium iodide solution, and 3 c.c. dilute sulphuric acid, starch solution being added when most of the liberated iodine has been decolourised. Care should be taken that the blue colour remaining in the solution due to the chromium salts is not confused with the blue starch iodide.

True N/40 Thiosulphate in c.c. $\times$ 0·056 = grains of Oxygen per gallon.

Grains of Oxygen per gallon $\times$ 9·964 = c.c. of Oxygen per litre.

Oil.—Extract 1,000 c.c. (or more), in successive quantities, with redistilled methylated ether in a globular separating funnel, more ether being added from time to time, if necessary, so that the ether layer is not less than 1 inch deep. After tapping off the last portion of the water, wash about three times with distilled water (about 10 c.c. each time), and finally distil off the ether in a tared flask (about 400 c.c. capacity), and dry to constant weight in a steam oven. If it is desired to include fatty acids present in the combined state as soap, the water should be acidified before extracting.

Weight of oil from 1,000 c.c. water, in grammes, $\times$ 70 = grains of oil per gallon.

Hardness by Soap Test.—The Standard Soap Solution used for this test may be either Clark's or Wanklyn's. If Clark's solution is used, 50 c.c. of water is taken. If Wanklyn's solution is used, 70 c.c. of water will be required.

In either case the required quantity of water is measured into a stoppered bottle having a capacity of about 200 to 250 c.c. The Soap Solution is then run in about 1 c.c. at a time, stoppering and shaking the bottle between each addition, until signs of a lather are noticed. Smaller quantities of the soap are then added, with shaking, till a good lather is formed which remains unbroken after standing for five minutes. It is better to lay the bottle on its side while waiting. In the presence of magnesia a false lather may be produced before the end point is reached, and this is sometimes mistaken for the real lather unless care is used. If there is any doubt about it, a reading of the burette should be taken, and another 0·5 c.c. of soap solution added and shaken. If the lather is the true one it will still remain, but a false one will probably have entirely disappeared. A further 0·5 c.c. may be added if there is still a doubt. If the lather disappears continue the titration.

In the case of Clark's solution, the Hardness corresponding to the number of cubic centimetres of soap solution used must be obtained by reference to " Clark's Table of Hardness " (see page 132). When Wanklyn's solution is used, 1 c.c. is deducted from the burette reading, and the remainder represents degrees of hardness.

Whichever solution is used, waters having more than 14 degrees of hardness must be diluted before testing, otherwise the results will be unreliable. Thus, if a water contains, say, 20 degrees of hardness, 25 c.c. may be diluted to 50 c.c. for the Clark test, and the figure for the hardness corresponding to the amount of soap solution used multiplied by 2 to ascertain the hardness of the original water. Similarly, if a water of 20 degrees of hardness is to be tested by the Wanklyn method, 50 c.c. can be diluted to 100 c.c., and 70 c.c. titrated. One cubic centimetre will then be deducted from the observed burette reading, and the remainder multiplied by 2, in order to obtain the hardness of the original water.

Calculation of Results.—It is impossible to say with certainty in what combinations the various bases and acid radicles found by analysis really exist. In all probability variations occur with changes of temperature and concentration, and any attempt to separate the various constituents is, therefore, rendered useless, as boiling and evaporation would be necessary, and the combinations thus materially changed. As it is desirable to have some idea as to what salts are likely to be present, or to be formed under certain conditions, some definite system must be adopted for the allocation of the acids to the various bases. When solutions of different salts are mixed, there is a tendency for the least soluble salts to be formed, and this may be taken as the best basis for the purpose of calculating the probable composition of a water.

The following instructions are given as a guide, but variations may at times be necessary, and must be made at the discretion of the analyst :—

Iron, Aluminium, Manganese, and Ammonia are probably nearly always present as Sulphates when actually in solution, and should, therefore, be usually first calculated as such.

All the Lime, or as much as possible (dependent on the amount of Alkalinity), should then be calculated to Calcium Carbonate.

Any remaining Lime, or as much as possible, to Calcium Sulphate.

Any remaining Lime, or as much as possible, to Calcium Nitrate.

Any remaining Lime, or as much as possible, to Calcium Chloride.

If any Alkalinity remains after calculating Calcium Carbonate, all the Magnesia, or as much as possible, should be calculated to Magnesium Carbonate.

Any remaining Magnesia, or as much as possible, to Magnesium Sulphate.

Any remaining Magnesia, or as much as possible, to Magnesium Nitrate.

Any remaining Magnesia, to Magnesium Chloride.

Any remaining Alkalinity or Acid Radicles may be calculated as Sodium Salts.

Silica may be expressed as such, or calculated as Sodium Silicate.

The difference between the Total Solids, and the sum of the calculated constituents, may be expressed as "Water not expelled at 270° Fahr., Organic Matter, etc." In some cases it may be mainly Organic Matter, in which case it will readily be noticed when the organic matter is being charred before the *lime* quantity is redissolved after evaporating to dryness.

Where Ammonium salts are present, the sum of the calculated constituents may exceed the estimated Total Solids, owing to the fact that all, or a portion of, the Ammonium salts will be expelled while the solids are drying.

Hardness is the sum of the Lime and Magnesia expressed in terms of Calcium Carbonate. If Iron, Aluminium, or Manganese are present they must also be calculated to Calcium Carbonate, and included in the Hardness, as they also have soap-destroying properties, but it is as well, in order to avoid any misunderstanding, to state "Hardness, including Iron, Aluminium, or Manganese."

"Scale-forming matter" may also be given, and should represent the Magnesia expressed as Magnesium Hydrate, plus Calcium Carbonate, Calcium Sulphate, and Silica. Iron, Aluminium, and Manganese should be included, if present, expressed as Oxides. Scale-forming matter may well be expressed in grains per gallon, and also in pounds per 1,000 gallons, the latter being one-seventh the former.

Calculation of Chemicals required for Softening.

Lime.—Calculate Lime equivalent to :—

(1) The Total Alkalinity (Alkalinity × 0·56).
(2) The Total Magnesia (MgO × 1·39).
(3) The Iron, Aluminium, and Manganese if present.

Add to the sum of these, Lime equivalent to the free CO_2 if it has been estimated ; or if it has not been determined, allow 1 grain of Lime per gallon, this being about equivalent to the average free CO_2. Divide the quantity of Lime thus calculated by 7 to bring it to pounds per 1,000 gallons, and then divided by 0·9 to obtain pounds of 90% Lime per 1,000 gallons. (Best Buxton Hard Lime is usually considered to contain about 90% of Caustic Lime.)

Soda-Ash (*Alkali* 58%).—Deduct the Total Alkalinity from the Hardness, and multiply by 1·08. This gives grains of Sodium Carbonate of 98% quality per gallon. Divide by 7 to bring it to pounds per 1,000 gallons.

Alkali 58% is the usual trade description of Soda Ash containing about 58% Na_2O, or 98% Na_2CO_3.

If the Alkalinity exceeds the Hardness, no soda ash is required, as sodium carbonate is already present in the water. If it is necessary to remove sodium carbonate from a water, Calcium Chloride or Alumino Ferric may be used, equivalent in quantity to the sodium carbonate present.

Calculation of Chemicals for the Treatment of Water when there is no Softening Plant.

(*a*) Deduct the Total Alkalinity from the Hardness, multiply by 1·08 and divide by 7. This gives the *equivalent* of Soda Ash in pounds per 1,000 gallons.

(*b*) If the Alkalinity exceeds, or is *not much less than*, half the Hardness due to Lime (total CaO × 1·784), only Caustic Soda will be required, in which case the Soda Ash calculated in paragraph (*a*) must be multiplied by 0·755 to obtain 98 per cent. Caustic Soda, and this alone should be used for the treatment.

(*c*) If the Alkalinity is *decidedly less than half* the Hardness due to Lime, some Soda Ash will be needed. Deduct twice the Alkalinity from the Total Lime Hardness, multiply by 1·08 and divide by 7.

This will give the Soda Ash, in pounds, required for the treatment of 1,000 *gallons of water.* Deduct this from the Soda Ash obtained in paragraph (*a*), and multiply the remainder by 0·755, to ascertain *the Caustic Soda necessary for* 1,000 *gallons of water.* A mixture of Caustic Soda and Soda Ash is used in this case.

CHAPTER X

ANALYSIS OF SCALE

As it is sometimes desirable to know the composition of the scale formed in boilers, the following brief description of a method of analysis may be useful :—

About 20 grammes of an average sample should be chosen, ground finely, and well mixed.

Water.—Weigh out 1 gramme into a platinum dish, dry at 300° Fahr. in an oven till constant in weight.

Organic Matter.—Heat the quantity used for water determination to faintest redness, till all organic matter is completely burnt, and weigh. The loss in weight will be due to organic matter (including oil) ; and combined water, if any, not expelled at 300° Fahr.

General Analysis.—Ignite 2 grammes very gently in a platinum dish to burn off organic matter. Cool, transfer to a beaker, and add a little bromine water to oxidise sulphides formed by the reduction of sulphates, and any ferrous salts. Cover the beaker, add hydrochloric acid, and boil till as much as possible of the scale has dissolved. Evaporate to dryness on a hot plate, and bake gently for half an hour. Cool, add 10 c.c. hydrochloric acid, heat for a short time, add about 100 c.c. of water, and heat till only silica remains undissolved. Filter into a 500 c.c. measuring flask, wash, dry, and ignite the silica (= Silica on 2 grammes).

Dilute the filtrate to 500 c.c., take 250 c.c. and estimate sulphates by evaporating to about 70 c.c., rinsing into a 6-oz. conical beaker, and precipitating with barium chloride as in water analysis (= Sulphates on 1 gramme).

Transfer 100 c.c. of the diluted filtrate from the silica to a 5-in. deep form porcelain basin, make just alkaline with ammonia, boil, and filter. Ignite and weigh the precipitate ; or separate iron, aluminium, and manganese as in water analysis (= Iron, etc., on 0·4 gramme).

Precipitate lime in the filtrate from the ammonia precipitate, with ammonium oxalate added 2 c.c. at a time till no further precipitate forms, filter, ignite, and weigh (= Lime on 0·4 gramme).

Slightly acidify the filtrate from the lime with hydrochloric acid, evaporate to about 70 c.c., make alkaline with ammonia, and precipitate Magnesia with sodium phosphate as in water analysis, filter, ignite, and weigh (= Magnesia on 0·4 gramme).

Carbonates.—It is not usually necessary to determine these, but if required the estimation may be carried out as follows :—

Place 1 gramme of the powdered sample in a flask of about 300 c.c. capacity, fitted with a cork having two holes. Through one hole insert a funnel with a stopcock, and having a neck so that a purifying train may be connected through the funnel to the flask. The stem of the funnel should reach nearly to the bottom of the flask, and should be drawn out to a small diameter at the end, and turned up for about $\frac{1}{4}$ in. The other hole in the cork is fitted with a short glass tube which is connected to the following series :—

(1) Dry, empty " U " tube.
(2) " U " tube of pumice soaked in concentrated sulphuric acid.

(3) Another "U" tube of pumice and sulphuric acid.

(4) "U" tube of pumice which has been soaked in a solution of copper sulphate, dried, and heated to render the copper sulphate anhydrous.

(5) Another "U" tube of sulphuric acid and pumice.

(6) Two "U" tubes tied together and connected in series, the first filled with soda-lime, and the second with pumice and sulphuric acid. These are the absorption tubes and should be weighed together.

In addition to these, a wash bottle of strong caustic soda solution, connected to a "U" tube of soda lime should be provided for the preliminary purification of the air required for driving the carbonic acid forward, and it must have a rubber tube by which the "U" tube can be connected to the top of the funnel in the evolution flask when necessary.

Put sufficient water in the flask to cover the sample, and insert the cork. Connect up to the long purifying train detailed above. Put 10 c.c. of hydrochloric acid in the stoppered funnel.

Commence by passing a current of air (about 1 litre per hour) through the preliminary purifying train.

Run the acid gradually from the funnel into the flask, but leave sufficient in the turned-up end of the stem to form a seal. When the effervescence has practically ceased, heat the contents of the flask, and boil gently for two minutes. Now connect up the preliminary purifying train to the funnel and open the stopcock, so that the current of purified air passes through the flask, and thus displaces the remaining carbonic acid. Turn out the flame under the flask. Allow the air to pass through the apparatus for half an hour, and then weigh the absorption tubes. The increase in weight = CO_2 on 1 gramme.

Oil.—Triturate 10 grammes of the dried scale (less if much oil be present), in a small mortar, with successive small quantities of redistilled methylated ether, decanting each portion through a dry filter into a tared flask (about 200 c.c.). Distil off the ether, and dry to constant weight in a steam oven.

Calculation of Results.—Calculate all the Sulphate to Calcium Sulphate. Then, if Carbonic Acid has been estimated, calculate it all (if possible) to Calcium Carbonate.

Should any Carbonic Acid remain, it must be calculated to Magnesium Carbonate.

If there is insufficient Carbonic Acid for the Lime remaining after calculating Calcium Sulphate, the excess Lime must be present as Calcium Hydrate (Caustic Lime).

Unless it has been necessary to calculate Magnesium Carbonate, the whole of the Magnesia should be expressed as Magnesium Hydrate.

If Carbonic Acid has not been estimated, any Lime left after calculating Calcium Sulphate should be expressed as Calcium Carbonate, and all the Magnesia as Hydrate.

If the sum of the calculated constituents, plus the Water and Organic Matter, exceeds 100 per cent., a sufficient quantity of the Magnesia must be expressed as Magnesium Oxide (instead of Hydrate) to reduce the total to 100 per cent.

CHAPTER XI

CONTROL TESTS FOR WATER SOFTENING

SOFTENING treatment should be carefully controlled by regular testing of the softened water. Tests ought to be made at least once a day, and where the water is liable to frequent changes in composition more frequent tests are necessary.

At the commencement, and from time to time afterwards, the plant should be carefully examined to ascertain whether it is working uniformly. In the case of tipping-buckets the number of tips per minute should be counted for several minutes, at a time when the demand for water is steady, and notice taken whether the buckets are filled to the same level each time.

All moving parts should be observed to detect any signs of sticking.

With plants which deliver the reagents in the form of solutions of constant strength, the strength of the solutions should be tested at intervals.

Any measuring devices should be cleaned, and kept clean.

The plant having been found apparently working satisfactorily, tests of the water as it enters and leaves the plant should be made, at intervals, several times during the day. Should the incoming water be found uniform in composition, but the treated water variable, the cause must be investigated and removed.

In most cases it is only necessary to determine the Hardness and Alkalinity. The method for Alkalinity described in Chapter IX. is quite simple, and can readily be made by any one, but it should be carried through carefully, especially the first part of the test where phenol phthalein is used as indicator.

Various methods have been devised for the determination of Hardness, but the soap test as described in Chapter IX., originally devised by Dr. Clark, is the most useful and convenient, and if carried out carefully usually gives sufficiently accurate results for the daily control of softening treatment. Methods necessitating boiling and evaporation are inconvenient, except in a chemical laboratory, and the results are not usually any more reliable than those obtained by the soap test, and some are decidedly unreliable.

Having made the tests, we must now consider what results are to be aimed at for a boiler feed water, and how they can be attained.

In the first place, the Hardness should be reduced as low as possible. It is frequently stated that this is undesirable, and that water for boiler feeding purposes should have at least 5 degrees of hardness. Experience does not confirm this, however, provided that other conditions are fulfilled. This mistaken idea has probably arisen partly owing to the fact that many of the naturally soft waters are more or less corrosive, and partly to the effects of some waters softened to a fairly low degree, but not fulfilling other conditions of efficient softening; or to softened waters used without sufficient regard to the effect of undue concentration, in boilers, of the soluble constituents.

Secondly, the Magnesia should be removed as completely as possible. Whereas Calcium Carbonate, which is probably the salt most easily removed by softening treatment, tends to neutralise the corrosive effect of other salts, Magnesium compounds are always apt to set up corrosive action. It will be clearly understood, therefore, that if imperfect softening results in the removal of most of the Calcium Carbonate while leaving the Magnesium salts, the corrosive tendencies of the latter have freer play; and whereas a water may have been used untreated without causing corrosion, inefficient softening may lead to wasting of the boiler plates, and the softening plant will be unjustly condemned.

The third important condition is that there should be a slight excess of Alkaline Sodium Salts in the treated water, unless other circumstances render then undesirable. Occasionally a portion of the softened water may be required for purposes other than boiler feeding, purposes for which an excess of soda would be harmful. In such cases it is better, if possible, to treat the water as for boiler feeding, and afterwards neutralise the portion required for other uses. It is also sometimes claimed that the steam from boilers fed with water treated with soda is objectionable for certain purposes, such as cooking. This can only be the case where either an undue accumulation of soda in the boiler is allowed, or where priming results in the carrying forward of the alkaline water with the steam.

To fulfil all the foregoing conditions, the Total Alkalinity must be *slightly higher* than the Hardness, and the Alkalinity of the water to Phenol Phthalein must be *slightly more than* half the Total Alkalinity.

If the Total Alkalinity is *less* than the Hardness, the proportion of Soda Ash requires increasing, while if it *exceeds* the Hardness by more than 0·5 degree it is usually desirable to reduce the Soda Ash. As every degree of Hardness or Alkalinity is equivalent to approximately $2\frac{1}{4}$ ounces of Soda Ash per 1,000 gallons, the proportion of Soda Ash must be increased or decreased by this amount for every degree difference between them, always allowing 1 ounce more Soda Ash per 1,000 gallons for the desirable excess.

If the Alkalinity to Phenol Phthalein exceeds *half* the Total Alkalinity, the presence of Caustic Alkalinity is indicated, and, as Magnesium Hydrate is almost insoluble in water, it follows that this condition is required for the effective removal of Magnesia. As a rule about 0·5 grain of Caustic Alkalinity is sufficient for the purpose, though with some waters more is desirable. If the N/50 acid required for the Alkalinity to Phenol Phthalein exceeds *half* that required for the Total Alkalinity by 1 c.c., 2 grains of Caustic Alkalinity is indicated, equivalent to 1·12 grains of pure Lime per gallon, or approximately 3 ounces of commercial Quick-lime of 90 per cent. quality per 1,000 gallons; and the proportion of Lime must therefore be increased or decreased by this amount for every 1 c.c. difference between the Alkalinity to Phenol Phthalein and *half* the Total Alkalinity, 1 ounce more Lime per 1,000 gallons being allowed for desirable excess.

The foregoing instructions are suitable for general guidance, but some waters may be found to require special conditions in order to obtain the best results. In such cases it is usually greater excess of reagents which is called for.

Having arrived at what appears to be satisfactory treatment, it is very desirable that a sample should be analysed for Lime, Magnesia, and Alkalinity in a laboratory, in order to ascertain whether Magnesia is being effectively removed, and also to act as a check on the soap test, and the accuracy of the standard solutions used in the daily testing. It is also advisable to have such a check carried out periodically, say once a month.

The standard solutions should also be tested occasionally, especially the soap solution, which, being made with alcohol, is very liable to evaporate and so become stronger. A Standard Sodium Carbonate solution of one-three hundred and fiftieth normal strength is useful for checking the Standard Acid, 70 c.c. of this strength requiring 10 c.c. of Standard Acid, and the test can be carried out with the same apparatus, and under the same conditions, as the water samples.

For checking the Standard Soap solution, a Standard Hard Water is needed. This may conveniently be of 14 degrees Hardness, and will be used exactly as in testing water samples.

If either solution is found to be appreciably too strong, it may be carefully diluted (using distilled water for the acid, or alcohol for the soap) till correct; or a factor may be made by which the volume of solution used in the titration may be corrected.

For works use it is usually considered better to have cylinders for the measurement of the water required for the tests, in preference to the more fragile pipettes.

CHAPTER XII

SAMPLING

Though it is a simple matter to collect a sample of water, the importance of fair sampling justifies a few remarks on the subject.

The most suitable bottles for the purpose are Winchester Quarts. They should have glass stoppers for preference, but if stoppered bottles cannot be obtained, clean sound corks may be used. Screw-stoppered bottles are unsuitable, as they have usually been used for ammonia, which permeates the screw stoppers and cannot be entirely removed by washing.

If possible, clear glass bottles should be obtained, so that the appearance of the sample can be judged without the necessity of transferring to another vessel.

The sample bottles, after thorough cleansing, should be rinsed out three times with the water to be sampled.

Whenever possible, samples should be collected directly into the bottles ; but if this is impossible, glass or earthenware vessels should be used in preference to those made of metal.

Bottles should always be filled to within an inch of the cork or stopper, but a little air space is essential, otherwise the bottle may burst.

Samples from ponds, tanks, or rivers should be taken by plunging the bottle (or other vessel) two inches or more below the surface of the water, in order to avoid any surface contamination.

When a sample is to be obtained from a tap, or a pump which has been standing, sufficient water should first be run to waste to avoid taking water which has been standing in the pipes.

If an estimation of dissolved gases is required, the sample should be collected with as little agitation as possible, the bottle filled as full as consistent with safety, and the sample kept as cool as possible until the determination has been made.

Boiler samples should be drawn from a water-gauge cock, after first flushing out. Care should be taken that the steam cock is closed, otherwise the sample will be diluted with condensed steam. In the case of boiler samples, a clean bucket may be used for collecting the hot water, which should then be poured into an earthenware vessel to cool somewhat before it is transferred to the bottle.

Every sample should be labelled with full particulars as soon as obtained, and the stopper should be covered with a piece of calico or other material, and securely tied down.

Winchester quart bottles can usually be sent by rail, per passenger train, without packing, with an addressed tag label tied to the neck. Except at holiday times samples thus sent usually reach their destination in safety. They are safer without packing than if badly packed.

If bottles are to be packed, the box should be large enough to allow plenty of clearance all round for wood-fibre or straw packing. No wedges should be used to hold the bottles in position, as this almost invariably results in breakage. Sawdust is not a good packing material, as it settles down into too solid a mass, and the bottles may break if the package receives a sudden jar.

CHAPTER XIII

SOLUTIONS

REAGENTS.

Hydrochloric Acid, sp. gr. 1·11.
HCl (1·16) 1,500 c.c., Water 900 c.c.

Nitric Acid, sp. gr. 1·2.
HNO_3 (1·42) 800 c.c., Water 1,280 c.c.

Ammonia, sp. gr. 0·95.
Ammonia (·880) 1,000 c.c., Water 1,500 c.c.

Washing Ammonia.
Ammonia (·880) 1 part, Water 3 parts.

Barium Chloride.
10 per cent. solution.

Ammonium Oxalate.
3·5 per cent. solution.

Sodium Phosphate.
Saturated solution of Disodium Hydrogen Phosphate.

Methyl Orange.
0·15 gramme per litre.

Phenol Phthalein.
3 per cent. solution in rectified spirit, decolourised if necessary by shaking with a little animal charcoal, filtered, and then neutralised by shaking with a little quicklime, and filtered.

Phenol Disulphonic Acid.

Dissolve 25 grammes of Phenol in 150 c.c. of concentrated sulphuric acid (1·84) and 75 c.c. fuming sulphuric acid (13 per cent. SO_3), and heat for 1 hour at 100° C.

Manganous Sulphate.

48 per cent. solution.

Sodium Hydrate and Potassium Iodide Solution.

360 grammes NaOH and 100 grammes KI dissolved in water and diluted to 1,000 c.c.

Nessler Solution.

Dissolve 62·5 grammes of KI in about 250 c.c. of distilled water. Reserve about 10 c.c. of this, and gradually add to the main portion a cold saturated solution of mercuric chloride until a slight permanent precipitate remains after shaking. Add potassium iodide and mercuric chloride solutions alternately till the former is nearly all used up, and a slight permanent precipitate remains. Add a cold solution of 150 grammes of caustic potash in 150 c.c. of water, and dilute to 1,000 c.c. Keep the solution in a rubber-corked bottle, decanting a portion of the clear solution into a smaller bottle from time to time as required.

Potassium Sulphocyanide.

10 per cent. solution.

Bromine Water.

Saturated solution in water.

Potassium Iodide.

10 per cent. solution.

Potassium Chromate.

10 per cent. solution, free from chlorides.

STANDARD SOLUTIONS.

Sulphuric Acid, N/50.

Dilute 10 c.c. of Normal Sulphuric Acid to 500 c.c.

Sodium Carbonate, N/350.

First make N/50 by diluting 10 c.c. of Normal Sodium Carbonate to 500 c.c.; and then dilute 71·4 c.c. of this to 500 c.c. (measure 70 c.c. with a pipette, and 1·4 c.c. from a burette).

Sodium Carbonate for CO_2 Determination.

1·723 grammes of *anhydrous* Sodium Carbonate per 500 c.c.

Silver Nitrate.

4·7944 grammes per litre (1 c.c. = ·001 gramme Chlorine).

Ammonium Chloride.

0·1568 gramme per litre (1 c.c. = 0·00005 gramme of NH_3).

Ferric Sulphate.

0·7 gramme of crystallised Ferrous Ammonium Sulphate and 5 c.c. dilute sulphuric acid, dissolved in water, made just pink with potassium permanganate, and diluted to 1,000 c.c. (1 c.c. = ·0001 gramme Fe).

Hard Water (14 degrees).

Dissolve 0·2 gramme of Calcite in dilute hydrochloric acid, and evaporate to dryness in a platinum dish. Add a little water and evaporate to dryness again, repeating this several times to remove all trace of acid. Dissolve in water, and dilute to 1,000 c.c.

Clark's Soap Solution.

Triturate 150 parts of Lead Plaster (Plumbi Emplast, B.P.) in a mortar with 40 parts of *anhydrous* Potassium Carbonate. When well mixed add a little methylated spirit, and again triturate until a smooth cream is formed. Add more spirit, stir, allow to stand for a short time, and filter. Keep this as a stock solution, diluting portion from time to time as required. Dilute with 2 parts of methylated spirit and 1 part water till 50 c.c. of Standard Hard Water requires 14·25 c.c. of the Soap Solution to produce a permanent lather.

Wanklyn's Soap Solution.

Dissolve 10 grammes of genuine Castile Soap in 1 litre of weak Alcohol (35 per cent. by volume). Filter, and dilute, if necessary, with weak Alcohol until 15·0 c.c. is required to produce a permanent lather with 70 c.c. of Standard Hard Water.

Sodium Thiosulphate.

6·2 grammes per litre, standardised against N/40 Potassium Bichromate.

Potassium Bichromate, N/40.

Heat powdered Potassium Bichromate over a very small flame till the colour just changes from orange to brown. Cool in a desiccator, and weigh out 0·6129 gramme. Dissolve in water and dilute to 500 c.c.

Potassium Nitrate.

0·3609 gramme per 500 c.c. (10 c.c. = ·001 grm. N).

APPENDIX

9

LIST OF FACTORS

Given.	Wanted.	Factor.	Given.	Wanted.	Factor.
Al_2O_3	$3CaO$	1·64589	Fe_2O_3	$2CaO$	·49015
Al_2O_3	$3SO_3$	2·35008	Mn_3O_4	$3MnSO_4$	1·97985
$BaSO_4$	SO_3	·34297	Mn_3O_4	$3SO_3$	1·04978
$BaSO_4$ grms. per 500 c.c.	SO_3 grs. per gal.	48·016	$Mg_2P_2O_7$	$2MgO$	·36225
CaO	$CaCO_3$	1·78482	$Mg_2P_2O_7$ grms. per 250 c.c.	MgO grs. per gal.	101·430
CaO mgrms. per 250 c.c.	$CaCO_3$ grs. per gal.	·5	MgO	CaO	1·39063
			MgO	$CaCO_3$	2·48202
			MgO	$MgCO_3$	2·09139
			MgO	$MgCl_2$	2·36210
CaO	Na_2CO_3	1·89058	MgO	Cl_2	1·75893
CaO	CO_2	·7848	MgO	SO_3	1·98562
CaO	SO_3	1·42788	MgO	N_2O_5	2·67907
CaO	$CaCl_2$	1·97946	MgO	$Mg(OH)_2$	1·44683
CaO	Cl_2	1·26485	MgO	Na_2CO_3	2·62909
CaO	N_2O_5	1·92652	MgO	$2NaOH$	1·98452
CaO	$2NaOH$	1·42708	$MgCO_3$	$CaCO_3$	1·18690
CaO	MgO	·71910	$MgSO_4$	Na_2CO_3	·88059
CaO	$MgCO_3$	1·50383	$Mg(NO_3)_2$	Na_2CO_3	·71461
$CaCO_3$	CaO	·56028	$MgCl_2$	Na_2CO_3	1·11303
$CaCO_3$	MgO	·40290	N_2	N_2O_5	3·85510
$CaCO_3$	$MgCO_3$	·84262	N_2O_5	Na_2O	·57397
$CaCO_3$	Na_2CO_3	1·05926	N_2O_5	MgO	·37326
$CaCO_3$	$2NaOH$	·79956	N_2O_5	CaO	·51907
$CaCO_3$	H_2SO_4	·98002	$2NH_3$	SO_3	2·35001
$CaSO_4$	Na_2CO_3	·77870	$2NH_3$	$(NH_4)_2SO_4$	3·87883
$Ca(NO_3)_2$	Na_2CO_3	·64602	SiO_2	Na_2O	1·02819
$CaCl_2$	Na_2CO_3	·95509	Na_2CO_3	$2NaOH$	·75482
CO_2	CaO	1·27417	Na_2CO_3	$CaCl_2$	1·04703
Cl	$NaCl$	1·64862	Na_2CO_3	CaO	·52894
Fe_2	Fe_2O_3	1·42980	Na_2CO_3	$BaCO_3$	1·86194
Fe_2O_3	$2CaO$	·70228	SO_3	Na_2O	·77442
Fe_2O_3	$2FeSO_4$	1·90256	SO_3	MgO	·50362
Fe_2O_3	$2SO_3$	1·00276	SO_3	CaO	·70026
Fe_2O_3	$2CaCO_3$	1·25344	H_2SO_4	SO_3	·81636
Fe_2O_3	$2FeCO_3$	1·45096	H_2SO_4	Na_2CO_3	1·08085
		·87482	H_2SO_4	CaO	·57170

ATOMIC WEIGHTS

(INTERNATIONAL, 1920)

Aluminium, . .	27·1	Manganese, . .	54·93
Barium, . .	137·37	Nitrogen, . .	14·01
Calcium, . .	40·07	Oxygen, . .	16·00
Carbon, . .	12·005	Potassium, . .	39·10
Chlorine, . .	35·46	Silicon, . . .	28·3
Hydrogen, . .	1·008	Silver, . . .	107·88
Iron, . . .	55·84	Sodium, . .	23·00
Magnesium, . .	24·32	Sulphur, . .	32·06

CLARK'S TABLE OF HARDNESSES

(In Degrees)

c.c.	0·0	0·1	0·2	0·3	0·4	0·5	0·6	0·7	0·8	0·9
0	..	..	..	..	..	..	..	0·00	0·11	0·22
1	0·33	0·44	0·55	0·67	0·78	0·89	1·00	1·09	1·18	1·27
2	1·36	1·45	1·54	1·64	1·73	1·83	1·93	2·01	2·09	2·18
3	2·28	2·37	2·46	2·55	2·64	2·73	2·82	2·91	3·00	3·10
4	3·20	3·30	3·40	3·50	3·60	3·70	3·80	3·90	4·00	4·10
5	4·20	4·30	4·40	4·50	4·60	4·70	4·80	4·90	5·00	5·10
6	5·20	5·30	5·40	5·50	5·60	5·70	5·80	5·90	6·00	6·10
7	6·20	6·30	6·40	6·50	6·60	6·70	6·80	6·90	7·00	7·11
8	7·21	7·32	7·42	7·53	7·63	7·74	7·84	7·94	8·05	8·16
9	8·26	8·37	8·48	8·58	8·69	8·79	8·90	9·00	9·11	9·21
10	9·32	9·42	9·53	9·63	9·74	9·84	9·95	10·06	10·16	10·28
11	10·39	10·50	10·61	10·72	10·83	10·94	11·05	11·17	11·28	11·39
12	11·50	11·61	11·72	11·83	11·94	12·05	12·17	12·28	12·39	12·50
13	12·61	12·72	12·83	12·94	13·06	13·17	13·28	13·39	13·50	13·61
14	13·72	13·83	13·94	14·06	14·17	14·28	14·39	14·50	14·61	14·72
15	14·83	14·95	15·06	15·18	15·29	15·41	15·53	15·65	15·76	15·88
16	16·00	..	..	..	..	..	..	..	..	..

Calculation of Saline Concentration in Boilers

Let C = number of gallons of water in boiler before blowing down.

D = ,, ,, ,, after ,,

S = saline matter in the feed water, in grains per gallon.

G = gallons of water evaporated per day (or per interval between blowings down).

Let $y = \frac{D}{C}$ (*i.e.*, fraction of total contents of boiler left after blowing down).

$x = \frac{G \times S}{C}$ (*i.e.*, increase in the saline matter in grains per gallon before blowing down, per day, or per interval between blowings down).

n = number of days (or intervals between blowings down).

Then—

Concentration, just before blowing down, after " n " days or periods (grains per gallon), . . . $= S\,y^{(n-1)} + \frac{x\,(1 - y^n)}{1 - y}$.

(The first term, $S\,y^{(n-1)}$, becomes so small after a few days' run that it can usually be omitted.)

Calculation of Days or Periods to reach a desired Concentration

Let A = desired concentration.

Then $n = \log \left\{ 1 - \frac{A\,(1 - y)}{x} \right\} \div \log y$.

Calculation of Maximum Possible Concentration

Let M = maximum possible concentration.

Then $M = \frac{x}{1 - y}$.

Calculation of Days or Periods to reach Maximum Possible Concentration

$$n = \log \frac{0{\cdot}5}{x} \div \log y.$$

CAPACITY OF HORIZONTAL CYLINDERS

Inches above Centre Line.	Diameter of Cylinders.													
	9′ 0″	8′ 6″	8′ 0″	7′ 6″	7′ 0″	6′ 6″	6′ 0″	5′ 6″	5′ 0″	4′ 6″	4′ 0″	3′ 6″	3′ 0″	2′ 6″
	Gallons per foot of length, at 6·24 gallons per cubic foot.													
1	4·68	4·42	4·16	3·90	3·64	3·38	3·12	2·86	2·60	2·34	2·08	1·82	1·56	1·30
2	9·36	8·84	8·32	7·80	7·28	6·76	6·24	5·72	5·20	4·68	4·16	3·63	3·11	2·59
3	14·03	13·25	12·47	11·70	10·91	10·13	9·35	8·57	7·79	7·01	6·22	5·44	4·66	3·87
4	18·70	17·66	16·62	15·58	14·54	13·50	12·45	11·41	10·37	9·33	8·28	7·24	6·19	5·14
5	23·37	22·06	20·76	19·46	18·16	16·85	1555	14·25	12·94	11·63	10·32	9·01	7·70	6·38
6	28·02	26·46	24·89	23·33	21·77	20·20	18·63	17·06	15·50	13·92	12·35	10·77	9·18	7·59
7	32·67	30·84	29·02	27·19	25·36	23·53	21·70	19·87	18·03	16·19	14·35	12·49	10·64	8·76
8	37·30	35·21	33·13	31·03	28·94	26·85	24·75	22·65	20·55	18·44	16·33	14·20	12·06	9·88
9	41·92	39·57	37·22	34·86	32·51	30·15	27·78	25·42	23·04	20·66	18·27	15·86	13·43	10·95
10	46·53	43·92	41·30	38·68	36·05	33·43	30·79	28·16	25·51	22·85	20·18	1749	1475	11·96
11	51·12	48·24	45·36	42·47	39·58	36·68	33·78	30·87	27·95	25·01	22·05	1906	16·02	12·89
12	55·69	52·55	49·39	46·24	43·08	39·91	36·73	33·55	30·35	27·13	23·88	20·58	17·22	13·72
13	60·25	56·83	53·41	49·99	46·55	43·11	39·66	36·19	32·71	29·20	25·65	22·05	18·34	14·44
14	64·78	61·10	57·40	53·71	50·00	46·28	42·55	38·80	35·03	31·23	27·37	23·44	19·37	15·00
15	69·29	65·33	61·37	57·40	53·42	49·42	45·41	41·37	37·31	33·20	29·03	24·76	20·30	15·31
16	73·77	69·54	65·31	61·06	56·80	52·52	48·22	43·90	39·53	35·12	30·62	25·98	21·09	
17	78·23	73·73	69·21	64·69	60·15	55·58	51·00	46·37	41·70	36·97	32·13	27·11	21·71	
18	82·65	77·88	73·09	68·28	63·45	58·61	53·72	48·80	43·82	38·75	33·55	28·11	22·05	
19	87·05	81·99	76·92	71·84	66·72	61·58	56·40	51·17	45·86	40·45	34·87	28·97		
20	91·43	86·08	80·73	75·35	69·95	64·51	59·02	53·47	47·84	42·07	36·09	29·65		
21	95·74	90·13	84·49	78·82	73·12	67·38	61·59	55·72	49·74	43·59	37·17	30·02		
22	100·04	94·13	88·21	82·25	76·25	70·20	64·09	57·88	51·55	45·01	38·09			
23	104·29	98·10	91·88	85·62	79·32	72·96	66·52	59·98	53·27	46·30	38·81			

24	108·50	102·02	95·51	88·95	82·34	75·66	68·89	61·98	54·88	47·45	39·21			
25	112·67	105·90	99·08	92·22	85·30	78·29	71·17	63·90	56·38	48·43				
26	116·80	109·73	102·61	95·44	88·19	80·85	73·38	65·71	57·75	49·20				
27	120·88	113·50	106·08	98·59	91·01	83·33	75·49	67·42	58·97	49·62				
28	124·90	117·23	109·49	101·67	93·76	85·72	77·50	69·00	60·01					
29	128·88	120·89	112·83	104·69	96·44	88·03	79·41	70·44	60·82					
30	132·80	124·50	116·12	107·64	99·03	90·24	81·19	71·72	61·26					
31	136·66	128·04	119·33	110·50	101·53	92·35	82·85	72·81						
32	140·46	131·52	122·47	113·29	103·94	94·34	84·36	73·66						
33	144·20	134·92	125·53	115·99	106·24	96·21	85·70	74·13						
34	147·87	138·25	128·51	118·59	108·44	97·94	86·84							
35	151·47	141·51	131·40	121·09	110·51	99·51	87·73							
36	155·00	144·68	134·20	123·49	112·46	100·91	88·22							
37	158·44	147·77	136·90	125·77	114·26	102·10								
38	161·81	150·76	139·50	127·93	115·90	103·02								
39	165·09	153·66	141·98	129·95	117·35	103·53								
40	168·29	156·46	144·34	131·81	118·58									
41	171·38	159·14	146·58	133·51	119·54									
42	174·37	161·71	148·67	135·02	120·07									
43	177·26	164·16	150·60	136·30										
44	180·03	166·46	152·36	137·29										
45	182·69	168·62	155·91	137·84										
46	185·21	170·62	155·25											
47	187·58	172·43	156·26											
48	189·81	174·04	156·83											
49	191·87	175·50												
50	193·74	176·46												
51	195·39	177·04												
52	196·80													
53	197·89													
54	198·49													

Note.—The table assumes the cylinders filled up to the horizontal centre line, and gives the complementary capacity, inch by inch, from the centre line upwards.

INDEX.

A

D

F

G

H

I

K

L

M

N

O

P

R

S

T

W

BELL AND BAIN, LIMITED, PRINTERS, GLASGOW.

Spotted something that isn't quite right?

Let us know and we'll fix it for future reprints. And as a thank you for your diligence, we'll send you a free Sothis book of your choice.

publisher@sothis.press

This page intentionally left blank

Except of course for that line above. And this one.
And whilst we have your attention, perhaps you'd
like to follow us on Instagram @sothis.press. Go on.

Made in the USA
Monee, IL
08 November 2026

67983954R00094